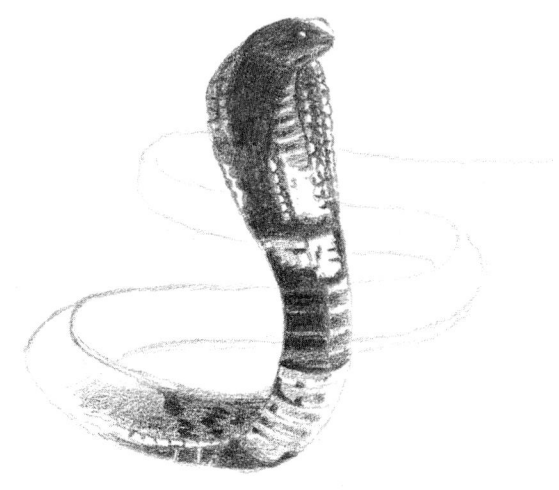

verlag josef margraf

AFRIKANISCHE IMPRESSIONEN
AFRICAN IMPRESSIONS

Helmut Poeschel

Zeichnungen und Text:
Helmut Poeschel
Gotthardt-Müller Str. 51
D-7024 Filderstadt

Übertragen ins Englische von:
Dr. Helmut Schmalfuss
Naturkundemuseum Stuttgart

Reproduktionen:
Ernst Müllerbader
Forststraße 18
7024 Filderstadt

Satz:
Trojanisches Pferd
Olgastraße 67
7000 Stuttgart 1

Druck:
Roth Offset
7311 Owen Teck

Bindung:
Idupa
In der Braike 21
7311 Owen Teck

1. Auflage 1985
© by Josef Margraf Verlag
Auf Aigen 3
7447 Aichtal

ISBN 3-924333-01-7

Umschlagbild:
(von links nach rechts)
Spießbock *(Oryx gazella gazella)*
Oryx-Antelope

Ringhals- oder Speikobra *(Haemachatus haemachatus)*
Rinkals Cobra

Kaffernhornrabe *(Bucorvus leadbeateri)*
Ground Hornbill

Kaffernbüffel *(Syncerus caffer caffer)*
Buffalo

Umschlagrückseite:
Tsavo-Elefanten *(Loxodonta africana)*
Tsavo Elephants

Meinen Angehörigen,
meinen Freunden
und all denen gewidmet,
die ihr Herz an Afrika verloren haben

To my family
my friends
and all who have lost
their hearts to
Africa

Schon als Junge begeisterten mich die Arbeiten der großen Künstler der Jahrhundertwende, wie die der Gebrüder Specht, Prof. Wilhelm Kuhnert, Prof. Richard Friese, um die wesentlichen zu nennen, welche es in der bildlichen Tierdarstellung zu einer meisterhaften Vollendung gebracht hatten. Das Buch »Im Lande meiner Modelle« von Prof. Wilhelm Kuhnert war meine Lieblingslektüre. Besonders hatten es mir Darstellungen über die afrikanische Tierwelt angetan. Sie erweckten in mir den Wunsch, diesen Kontinent mit seiner vielseitigen und interessanten Fauna und Flora selbst kennen zu lernen und zu erleben.

Seit dieser Zeit haben sich in Afrika große Veränderungen ergeben. Der Tierreichtum des Kontinents beschränkt sich inzwischen im wesentlichen auf die Nationalparks und Schutzgebiete. Trotzdem kann man heute noch »Afrika erleben«. Eine große Zahl afrikanischer Staaten macht enorme Anstrengungen, die charakteristische Tier- und Pflanzenwelt jeweiliger Gebiete zu erhalten. Die persönliche Begegnung mit Elefanten, Löwen und großen Herden von Antilopen und Zebras ist für jeden Menschen, der sie erfahren hat, ein Lebenshöhepunkt.

Nachdem ich verschiedene afrikanische Länder bereist und ihre Tierwelt studiert hatte, entstand der Wunsch, die gesammelten Eindrücke in einem Buch zu veröffentlichen. Es gibt viele hervorragende Farbbildbände über Afrika, seine Landschaften, Tiere und Pflanzen. Vielleicht ist gerade deshalb eine nur mit dem Bleistift gestaltete Darstellung reizvoll. Jedes Tier hat ein für seine Art charakteristisches Verhalten, ob bei der Nahrungsaufnahme, am Wasser, im Schreckmoment oder auf der Flucht, an der Silhouette ist es zu erkennen.

Vielleicht trägt dieses Buch auch zu aufmerksamerer und intensiverer Natur-und Tierbeobachtung bei, um eine »Safari« zum großen Erlebnis werden zu lassen. Mögen uns allen diese herrlichen Naturgebiete durch ein verständiges und einfühlungsvolles Benehmen gegenüber der Schöpfung erhalten bleiben.

Dem Verleger Josef Margraf, der mir volle Freiheit bei der Gestaltung des Buches ermöglichte, meiner Familie und allen Freunden, welche mich auf meinen Afrika-Reisen begleiteten oder mir in oft schwierigen Situationen Unterstützung zukommen ließen, sei an dieser Stelle herzlich gedankt.

Filderstadt, im Frühjahr 1985　　　　　　　　　　　　　　　　　　　　Helmut Poeschel

When I was a child I was already fascinated by the works of those great artists of the turn of the century who had achieved perfection in portraying animal life, the brothers Specht, Professor Wilhelm Kuhnert, Professor Richard Friese, to name the most outstanding of them. The book »Im Lande meiner Modelle« (In the land of my models) by Wilhelm Kuhnert was my favorite reading. I was especially enchanted by pictures of African animal life. This inspired my desire to know and to experience this continent with its diverse and interesting fauna and flora.

Since that time Africa has undergone great changes. The richness of animal life has been restricted to the national parks and game reserves. Yet it is still possible to »experience Africa«. A great number of African states endeavor with great efforts to encourage the survival of their characteristic animal and plant life. Personal encounters with elephants, lions and big herds of antilopes and zebras are highlights in the life of anyone who has witnessed such spectacular scenes.

After I had travelled in a number of African countries, studying their animal life, and portraying it in many pencil drawings, I wished to publish a collection of these impressions. There are many excellent publications of color photographs on Africa, its landscapes, its animals and plants. Perhaps a presentation only with pencil drawings will be rewarding as a contrast.

May this book contribute to a more attentive and a more intensive engagement with nature and animals, for a »safari« to become a really great experience. May these wonderful nature reserves be saved for all of us by an appreciative behavior of mankind towards Creation.

I wish to thank the publisher Josef Margraf, to whom I owe the opportunity to realize my personal conceptions of the book, and my family and all friends, who were my companions on my journeys through Africa, or who gave me their support in difficult situations.

Filderstadt, spring 1985　　　　　　　　　　　　　　　　　　　　　　Helmut Poeschel

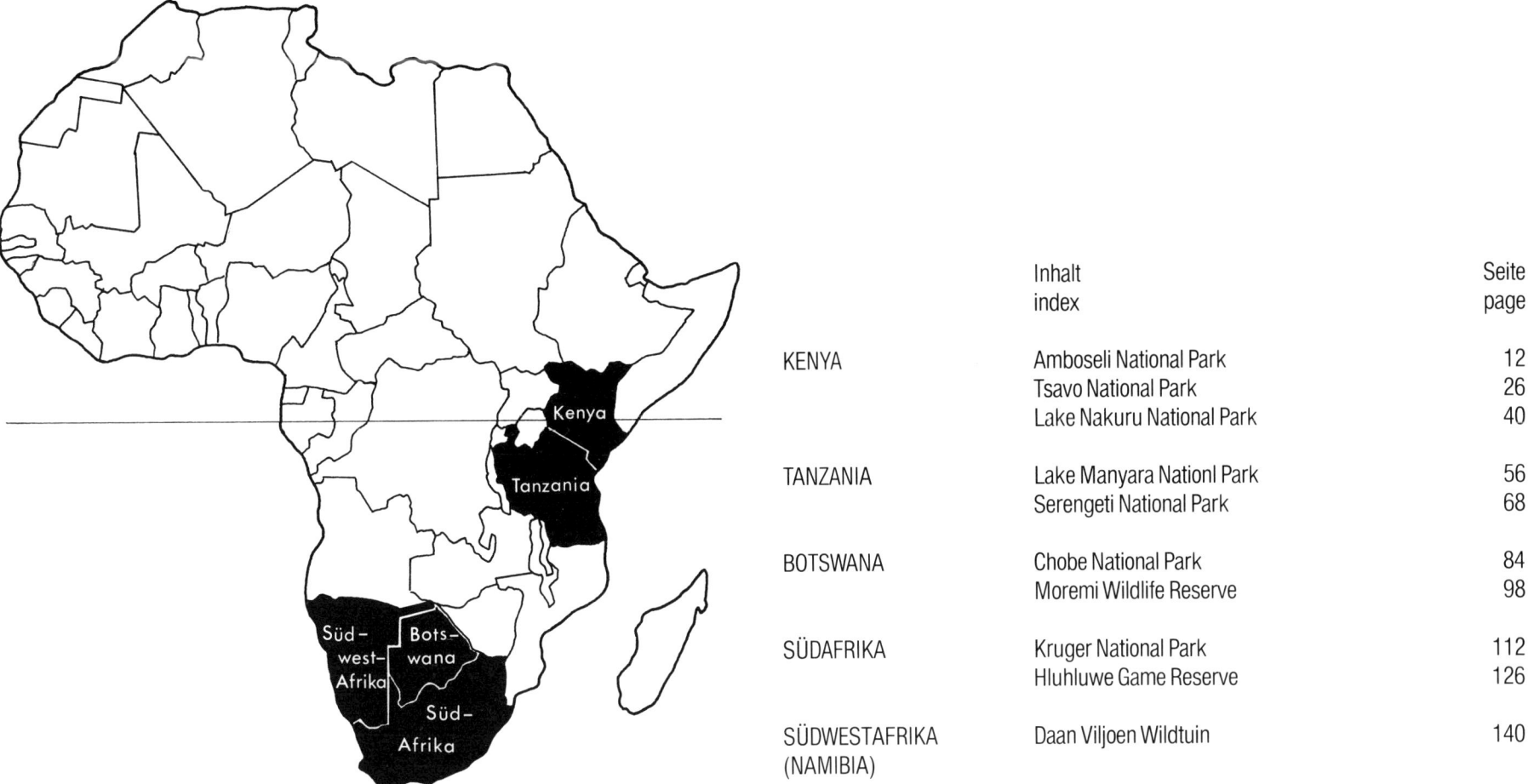

	Inhalt index	Seite page
KENYA	Amboseli National Park	12
	Tsavo National Park	26
	Lake Nakuru National Park	40
TANZANIA	Lake Manyara Nationl Park	56
	Serengeti National Park	68
BOTSWANA	Chobe National Park	84
	Moremi Wildlife Reserve	98
SÜDAFRIKA	Kruger National Park	112
	Hluhluwe Game Reserve	126
SÜDWESTAFRIKA (NAMIBIA)	Daan Viljoen Wildtuin	140

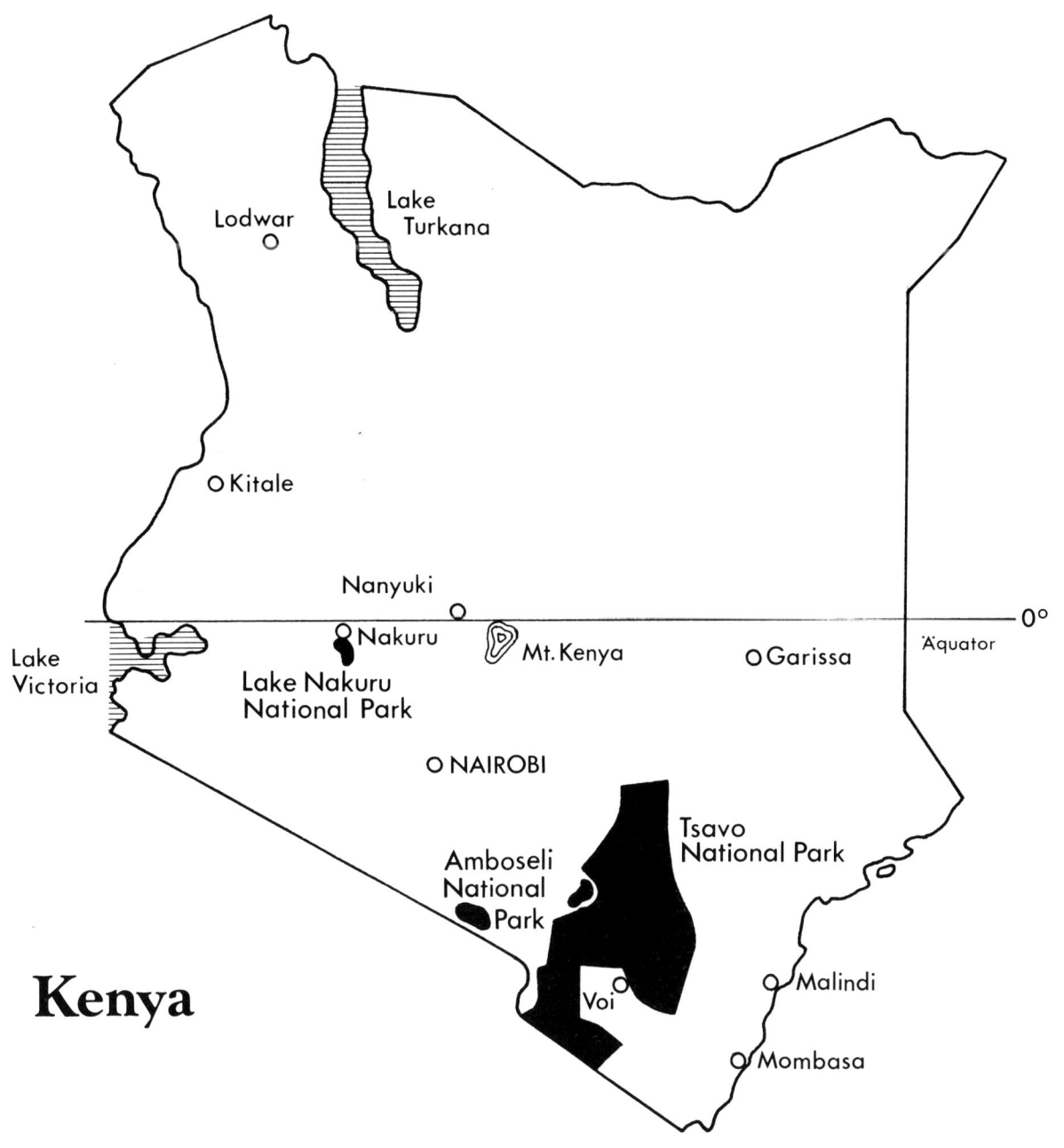

Aus dem unter Verwaltung der Kenya-Nationalpark-Behörde stehenden, im Jahre 1960 gegründeten Nationalreservat entstand 1975 der Amboseli-Nationalpark. Das ursprünglich 3200 qkm große Gebiet wurde auf ein Bruchteil von 150 qkm reduziert. Der 1170 m über dem Meeresspiegel liegende Nationalpark wird von der majestätischen Kulisse des in Nord-Tanzania gelegenen 5895 m hohen Kilimanjaro überragt.

Die Landschaftsformen gliedern sich in große Grasflächen, Sumpfgebiete und kleinere Seen, in Wälder aus Gelb- und Schirmakazien und in trockenes Buschland. Der Amboseli-See ist nur bei besonders reichlichen Niederschlägen teilweise mit Wasser bedeckt, den übrigen Teil des Jahres bildet er eine trockene alkalische Salzfläche. Der Boden ist Vulkan-Asche aus der früheren Aktivität des Kilimanjaro-Massivs.

Der Amboseli-Nationalpark beherbergt eine große Artenvielfalt an typischen Savannentieren, welche man auf verhältnismäßig kleinem Raum beobachten kann. Wildtierkonzentrationen sind immer jahreszeitlichen Gegebenheiten angepaßt, wobei das Nahrungsangebot entscheidend ist. So wandern auch im Amboseli gewisse Tierarten und größere Herden ab. Ein Besuch zu diesem Zeitpunkt kann dann enttäuschend sein. Ich habe Amboseli in beiden Extremen erlebt. Gute Tierbeobachtungsmöglichkeiten sind bei Ol Tukai und im Gebiet der Loginya-Seen.

Der Amboseli-Park dürfte wohl in Kenya die höchsten Besucherzahlen aufweisen und es ist keine Seltenheit, daß eine schlafende Löwengruppe oder eine Gepardin mit Jungen von Kleinbussen regelrecht eingekreist und sogar bei der Jagd behindert wird. Negativ wirkt sich auch die Überweidung durch die Viehherden der Massai bis in das Kerngebiet des Parks aus. Dies stellt eine ernste Konkurrenz für die Wildtiere dar und eine Gefahr für den Fortbestand des Reservats. Trotzdem gibt es für den ernsthaften Naturfreund abseits der von Eiltouristen befahrenen Hauptwege Parkgebiete, in welchen man ungestört und mit viel Muse die reichhaltige Tierwelt erleben und beobachten kann.

A greatly reduced part of the area originally instituted as a national reserve in 1960 under the administration of the Kenyan National Park authorities was declared the Amboseli National Park in 1975.

Originally covering 3,200 square kilometers the area included today in the Amboseli Park consists only of 150 square kilometers. This park is situated at an elevation of 1170 meters and is towered by the majestic scenery of the Kilimanjaro in North Tanzania, which rises up to 5895 meters.

The landscape formations contain extensive grasslands, marshy areas with small lakes, acacia stands and dry bush. Thc Lakc Amboseli is covered with water only in rainy periods, the remaining part of the year it is a dry flat of crystallized salt. The ground consists of volcanic ashes, the result of the former activity of the Kilimanjaro.

The Amboseli Park is populated by a great variety of typical savannah animals, which can be observed in a relatively small area. Concentrations of wild animals are always dependent on seasonal conditions, the food supply being the most important factor. Some game species and greater herds leave the Amboseli Park at certain times of the year, so a visit can sometimes be very disappointing. I have seen the Amboseli Park at both extremes. Good possibilities to watch animals are at Ol Tukai and in the region of the Loginya Lakes.

The Amboseli Park has the highest number of visitors of the Kenyan parks, and it often happens that a sleeping pack of lions or a female cheetah with cubs is completely encircled by minibusses. Another negative aspect is the over-grazing by Massai cattle right to the central areas of the park. This means serious competition for wild game and endangers the further existence of the park.

Despite these negative aspects there are areas in the park off the main routes where the abundant animal life can be watched undisturbed.

Amboseli National Park Kenya

Der Tsavo-Nationalpark ist mit seinen 20600 qkm das größte Wildschutzgebiet Kenyas. Der Park wird durch die Straße Nairobi-Mombasa in Tsavo-West und -Ost geteilt. Durch Tsetse-Verseuchung und Wassermangel war das Gebiet nicht zur Besiedlung geeignet. Der ungewöhnliche Tierreichtum veranlaßte schon die englischen Kolonialbehörden, ein Jagdverbot zu erteilen und das Gebiet später unter Schutz zu stellen.

Der Park beinhaltet verschiedene Lebensräume, vorwiegend offenes Savannen-Buschland mit Dornbüschen und Akazienwäldern. Besonders beeindruckend sind die imposanten Kulissen erloschener Vulkanberge. Im Tsavo-West führt der Tsavo Fluß ganzjährig Wasser, im Tsavo-Ost ist es der durch Zusammenfluß von Athi und Tsavo gebildete Galana. Der vulkanische Charakter ist im Westteil vorherrschend, im Ostteil sind es Gneis- und Schieferformationen.

Ein besonders idyllischer Platz sind im Tsavo-West die Mzima-Springs. Hier sprudelt klares Quellwasser aus dem Lavagestein und bildet einen See, in welchem Flußpferde und vereinzelt auch Krokodile leben. Die Hauptanziehungskraft für den Besucher sind jedoch die ständig zum Wasser kommenden Tierherden. Besonders die Elefanten sind es, die stets in kleineren und größeren Herden am Wasser anzutreffen sind und ein imposantes Schauspiel abgeben. Mzima war noch vor einigen Jahren ein stiller und ruhiger Platz, heute werden hier sämtliche Touristen auf ihrer berühmten »Drei-Tage-Safari« durchgeschleust.

Tsavo ist Elefantenland und man schätzt den Bestand auf etwa 18000 Tiere. Überall begegnet man den von einer Leitkuh angeführten Familien- oder Herdenverbänden. Aber auch einzelne oder mehrere Bullen mit starken Zähnen, überpudert mit roter Laterit-Erde, kreuzen den Weg. Durch die hohe Population wurde die Vegetation sehr in Mitleidenschaft gezogen, vor allem die alten Baobabs sind oft übel zugerichtet oder schon ganz zerstört. Die Wilderei ist im Tsavo ein besonderes Problem, ihr sind schon viele Elefanten und vor allem Nashörner zum Opfer gefallen. Wenngleich im Tsavo keine größeren Tierherden zu beobachten sind, hat man doch ständig Begegnungen mit einer artenreichen Tier- und Vogelwelt.

Der Tsavo ist für mich einer der schönsten und beeindruckensten Nationalparks. Die Weite der Landschaft, die Ruhe, die Unerschöpflichkeit an Motiven in Form und Farbe sind ein Stück Afrika, an das ich mein Herz verloren habe.

The Tsavo National Park is the largest game reserve of Kenya, extending over 20,600 square kilometers. The park is divided into an eastern and a western part by the road Nairobi - Mombasa. Contamination by tsetse-flies and lack of water have prevented human settlements. The unusual abundance of game had already induced the English colonial authorities to prohibit hunting, and later on to declare it a protected area.

The park consists mainly of open savannah and bush with thorny scrub and acacia stands. The extinct volcano mountains make for an especially impressive scenery. In the western part the Tsavo river has perennial water, in the eastern part the joined Tsavo and Athi form the Galana river. The volcanic character is predominant in west Tsavo, while in the east there are gneiss and slate formations.

An exceptionally idyllic place in Tsavo West is Mzima Springs. Clear water gushes out of the lava rock, creating a lake where hippos and crocodiles live. The main attraction for the visitor are, however, the animals continuously coming to the water to drink. Elephants can be seen there at all times in smaller or bigger herds. A few years ago Mzima was a quiet spot, today all the tourists of the famous three-days-safaris are brought to this place.

Tsavo is elephant's land, their number is estimated 18,000. Everywhere in the park the family herds, led by an experienced cow, can be met with, as well as single bulls or small groups of bulls, often with considerable tusks, and powdered red with laterite dust. The vegetation has been badly affected by this high population, especially the old baobab trees have been damaged or completely destroyed. Poaching is a serious problem in the Tsavo Park, concerning elephant and rhino.

The enormous herds of game are missing in the Tsavo, but a great number of mammal and bird species can be encountered there. For me the Tsavo is one of the most beautiful and impressive national parks. The vastness of the landscape, the silence, the inexhaustible richness of motifs in form and color are the fascination of this piece of Africa which I have become very attached to.

Tsavo National Park
Kenya

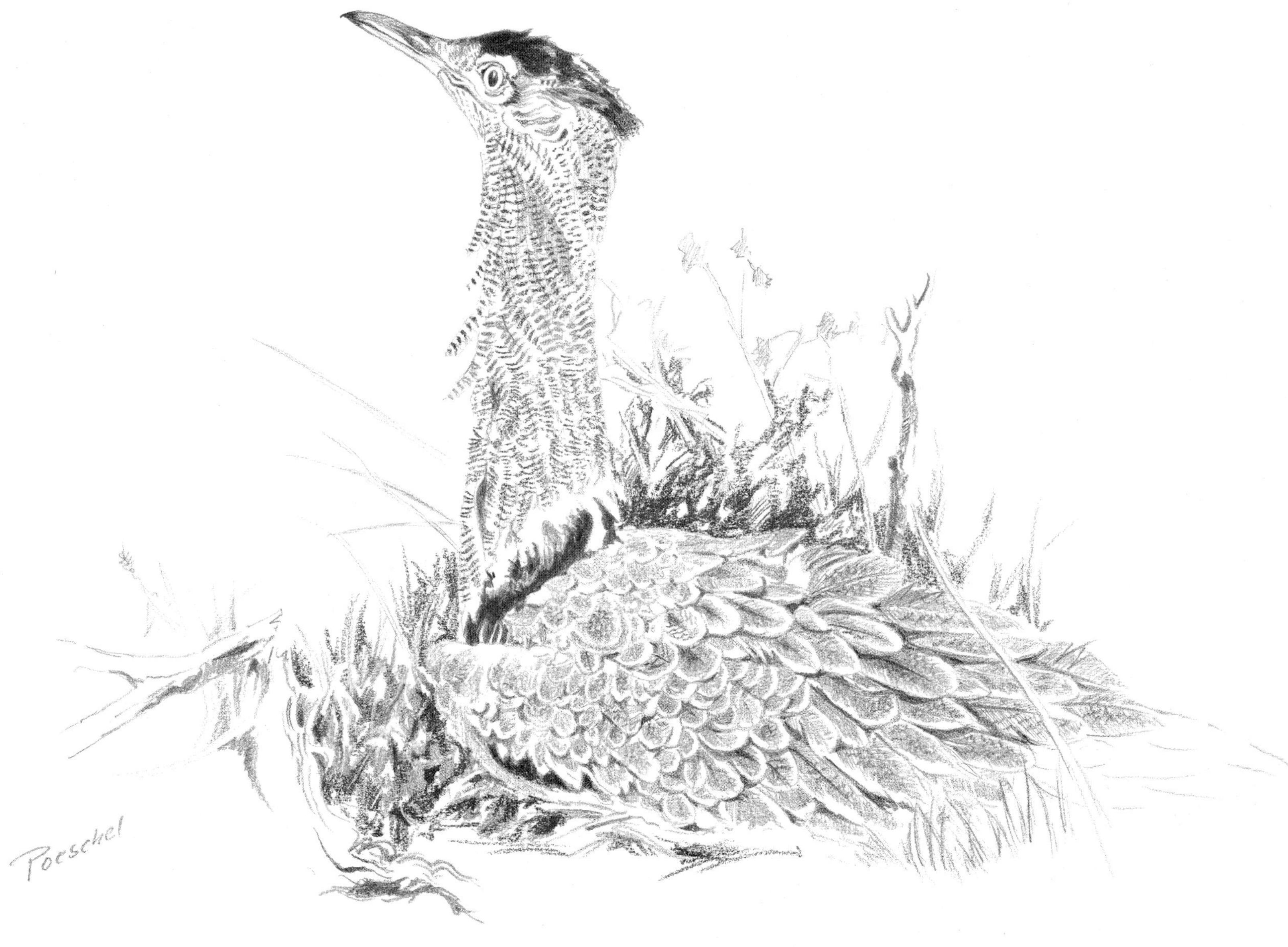
Poeschel

36

Poeschel

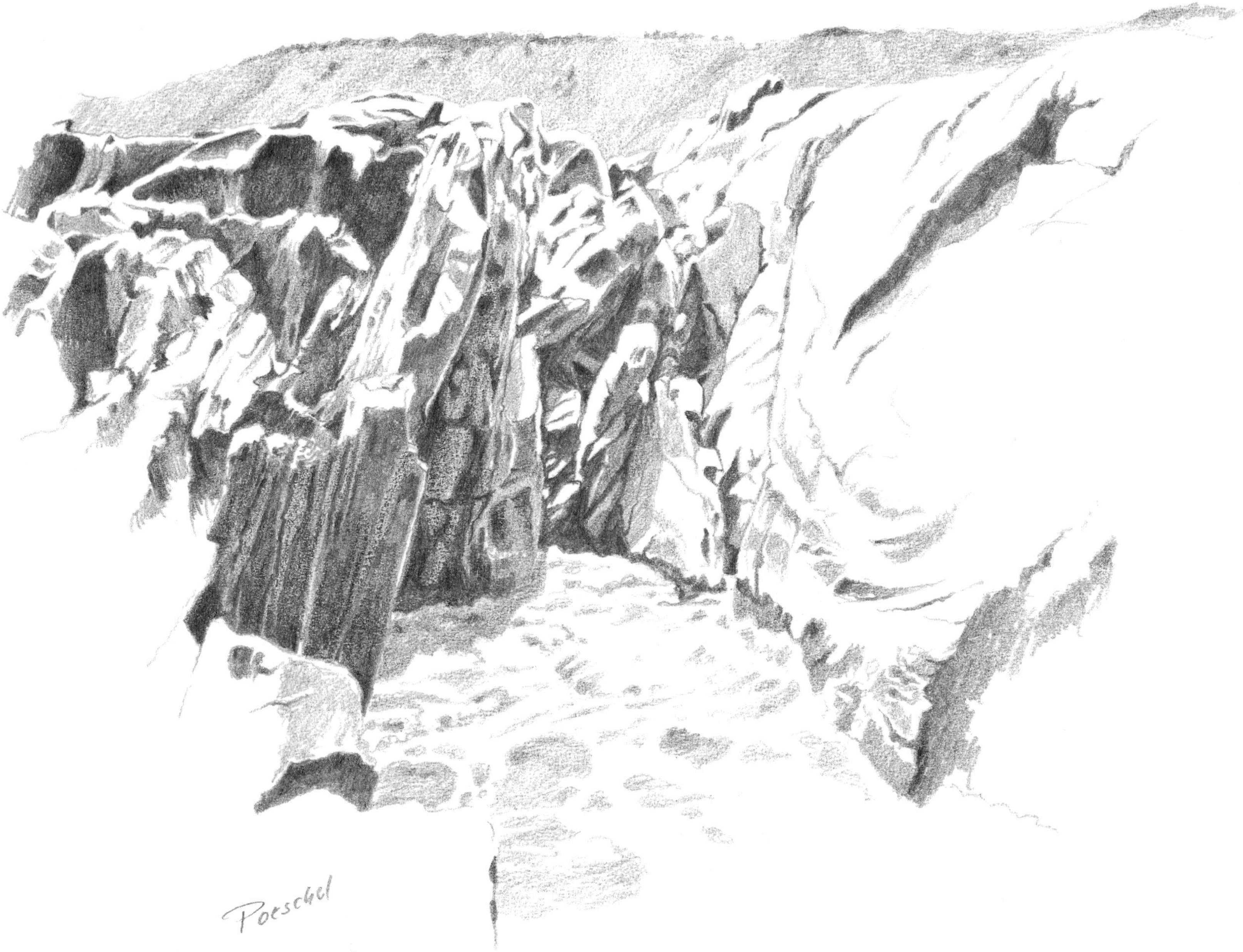
Poeschel

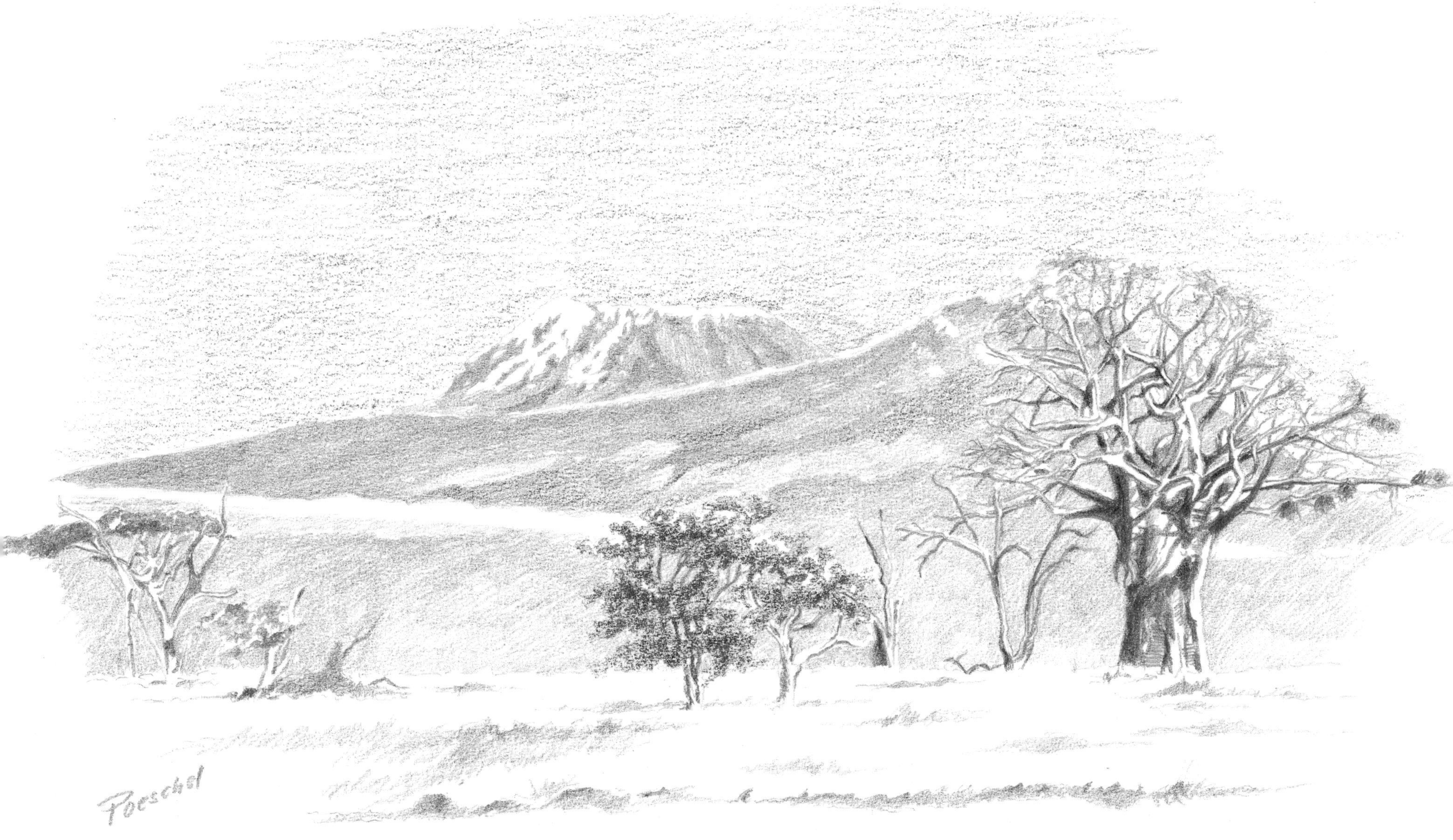
Poeschel

Zu den Besonderheiten des ostafrikanischen Grabenbruchs gehören die verschiedenen sodahaltigen Seen, von welchen der Nakuru-See besonders bekannt wurde. Wegen seiner zeitweilig dort lebenden 1 bis 2 Millionen Flamingos bezeichnete ihn der amerikanische Ornithologe Roger Tory Peterson »als größtes Wunder der Vogelwelt«.

1961 wurde ein Teil des südlichen Sees zum Nationalpark erklärt, ihm folgte 1968 der nördliche Teil. Ende der 70er Jahre wurde er durch weitere Gebietsanschlüsse vergrößert, sodaß heute der ganze See sowie das anschließende Wald- und Buschland zum Nationalpark gehören. Ein gutes Wegenetz wurde angelegt, und eine Lodge sowie mehrere Campingplätze werden allen Besucherwünschen gerecht.

Der See grenzt unmittelbar an die Stadt Nakuru und ist etwa 60 qkm groß. Die von der Stadt in den See geleiteten Abwässer stellen inzwischen eine ernste Bedrohung für die Existenz der Vogelwelt dar. Der World Wildlife Fund sowie andere Institutionen halfen bereits mit Geldmitteln. Auch die Bundesregierung sprang mit einer großzügigen Spende ein, die den Bau einer Abwasseraufbereitungsanlage ermöglichte.

Die Landschaft ist sehr abwechslungsreich. Sie umfaßt Schilf-, Sumpf- und Grasgebiete mit Felsklippen, Wälder gelbrindiger Akazien, sowie die Baboon-Rocks, eine Felskette am westlichen Seeufer und dem Lion Hill im Osten. Besonders erwähnenswert ist ein Wald aus sonst nur vereinzelt in der Savanne vorkommenden Kandelaber-Euphorbien (Euphorbia ingens). Der Park wurde vor allem zum Erhalt der Vogelwelt gegründet und bildet ein unerschöpfliches Dorado für Ornithologen und Vogelfreunde. Der Einstand der Flamingos ist vom Nahrungsangebot und damit von der Höhe der Niederschläge abhängig. Ich selbst habe bei meinen Besuchen nur kleine Gruppen zu Gesicht bekommen. Wasser- und Watvögel wie auch uferbewohnende Arten sind überall in beeindruckender Zahl zu sehen, und auch die Säuger sind artenreich vorhanden.

Ich hatte hier Gelegenheit, meinen ersten Leoparden in Afrika zu sehen und die Gruppe der seltenen Colobus-Affen im Gelbakazienwald am südlichen Teil des Sees. Für den Nakuru-Nationalpark muß der Besucher viel Zeit mitbringen, um all die vielen großen und kleinen Naturwunder in sich aufnehmen zu können.

The East African Rift Valley contains a number of soda lakes, of which Lake Nakuru is especially famous. At times there are one to two million flamingos present on this lake, and the American ornithologist Roger Tory Peterson called it the greatest ornithological miracle of the world.

Part of the southern lake was declared a national park in 1961, followed by the northern part in 1968. In the seventies further areas were included, so that today the complete lake as well as adjacent forest and bush areas belong to the national park. Good roads were constructed, a lodge and several campsites guarantee convenience for the visitor.

The lake extends over about 60 square kilometers directly adjacent to the town Nakuru. The town's sewage being drained straightaway into the lake became a serious threat for the birdlife. The World Wildlife Fund and other institutions helped with money. Finally it was possible to build a sewage farm with money generously donated by the German government.

The landscape varies considerably. There are reed beds, marshy areas, grasslands, rocky cliffs, acacia woods, the Baboon Rocks, a rocky chain on the western lake shore, and the Lion Hill on the eastern side. Especially remarkable is a dense wood of a succulent euphorbia (Euphorbia ingens) which is elsewhere found only in single specimens in the savannah. The park was established mainly for the conservation of the birdlife and is an inexhaustible dorado for ornithologists and birdwatchers. The number of flamingos depends on the food supply, this being determined by the height of the water level and thus by the extent of rainfall. I have seen only small parties of flamingos during my visits. Waterfowl, waders, and species living at the shore are always met with in impressive numbers.

Here I saw my first leopard in Africa, and the group of rare Colobus Monkeys in the acacia forest south of the lake. The visitor of the Nakuru Park should not be in a hurry, to absorb and appreciate all the great and small wonders of nature.

Lake Nakuru National Park
Kenya

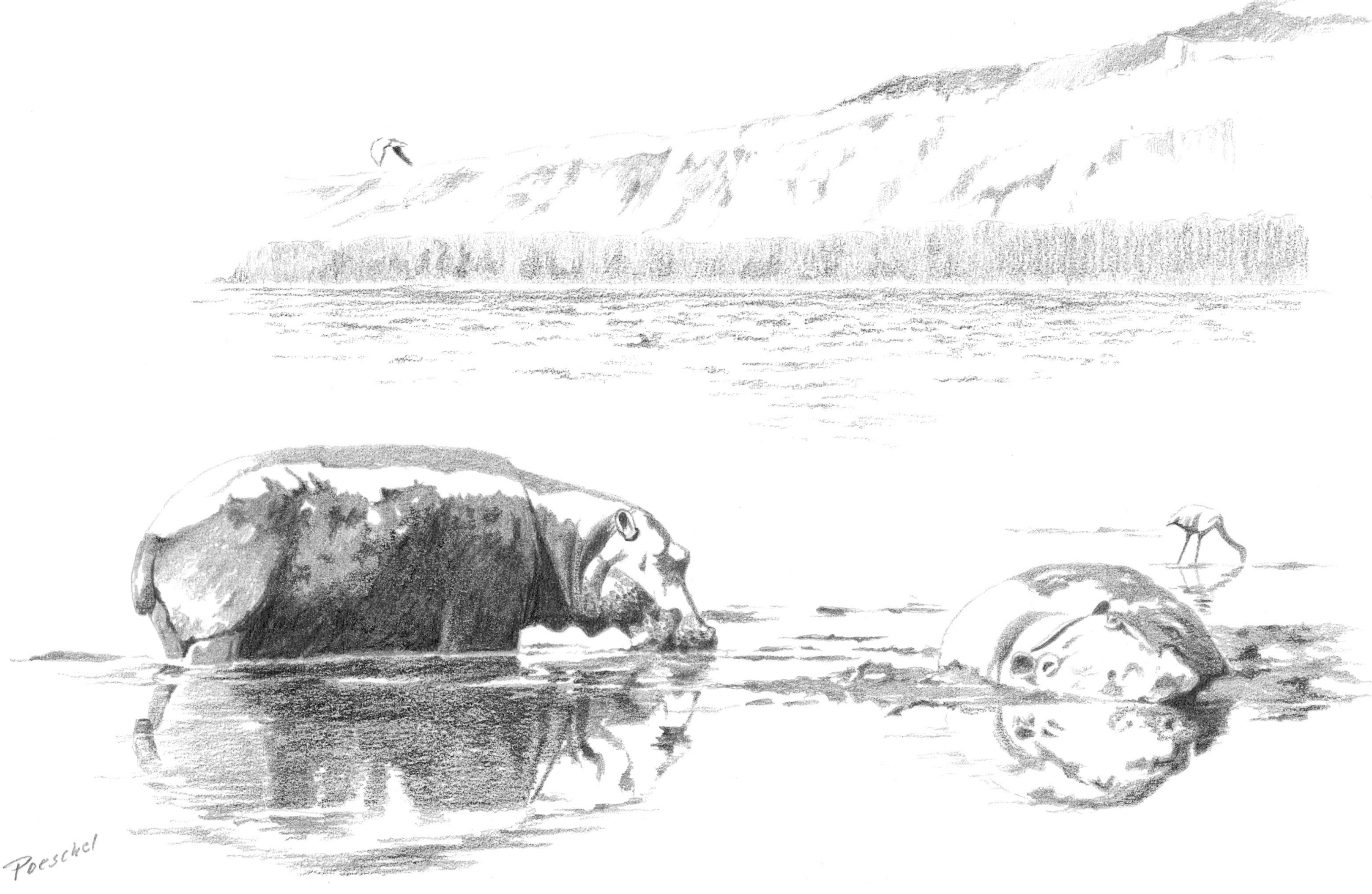
Poeschel

43

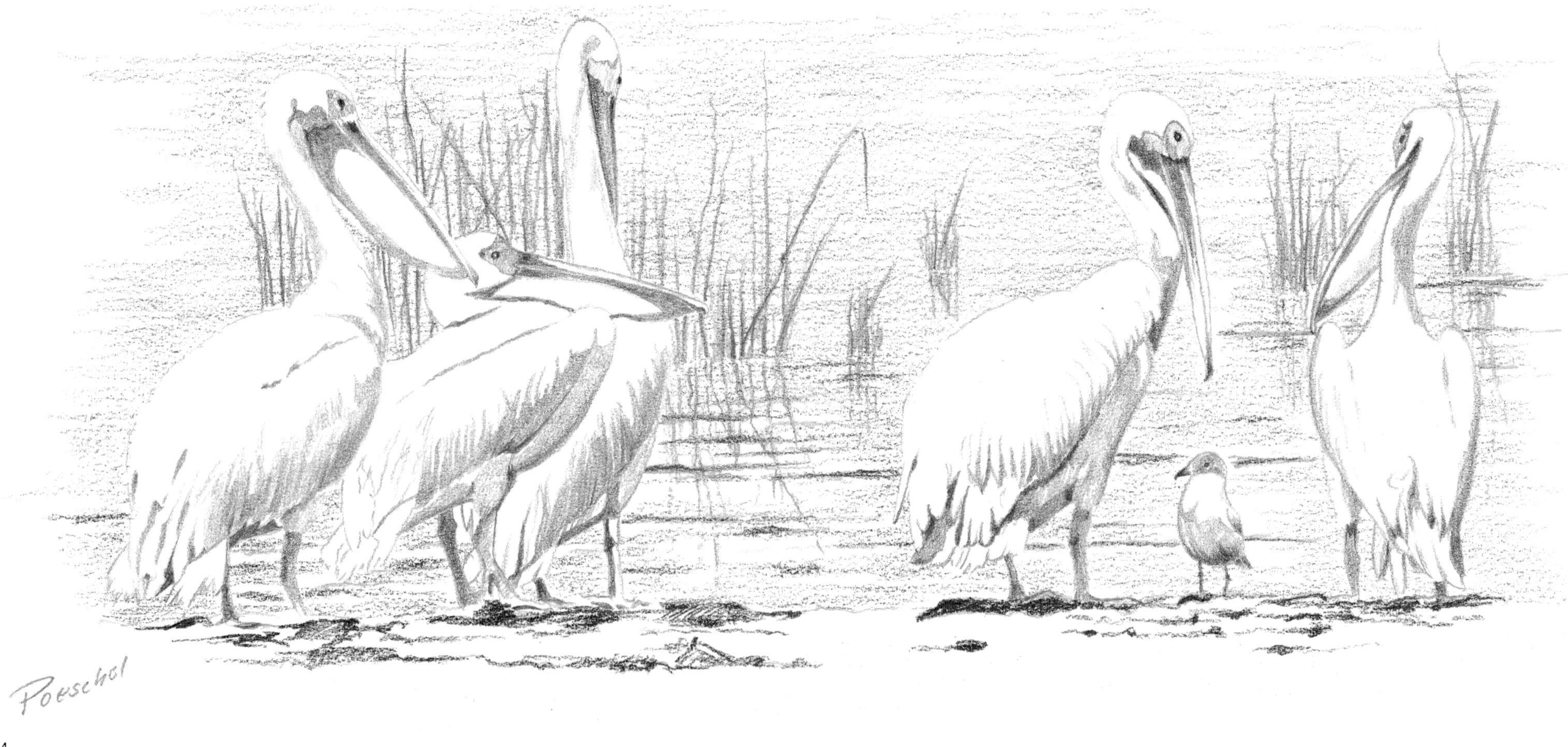
Poeschel

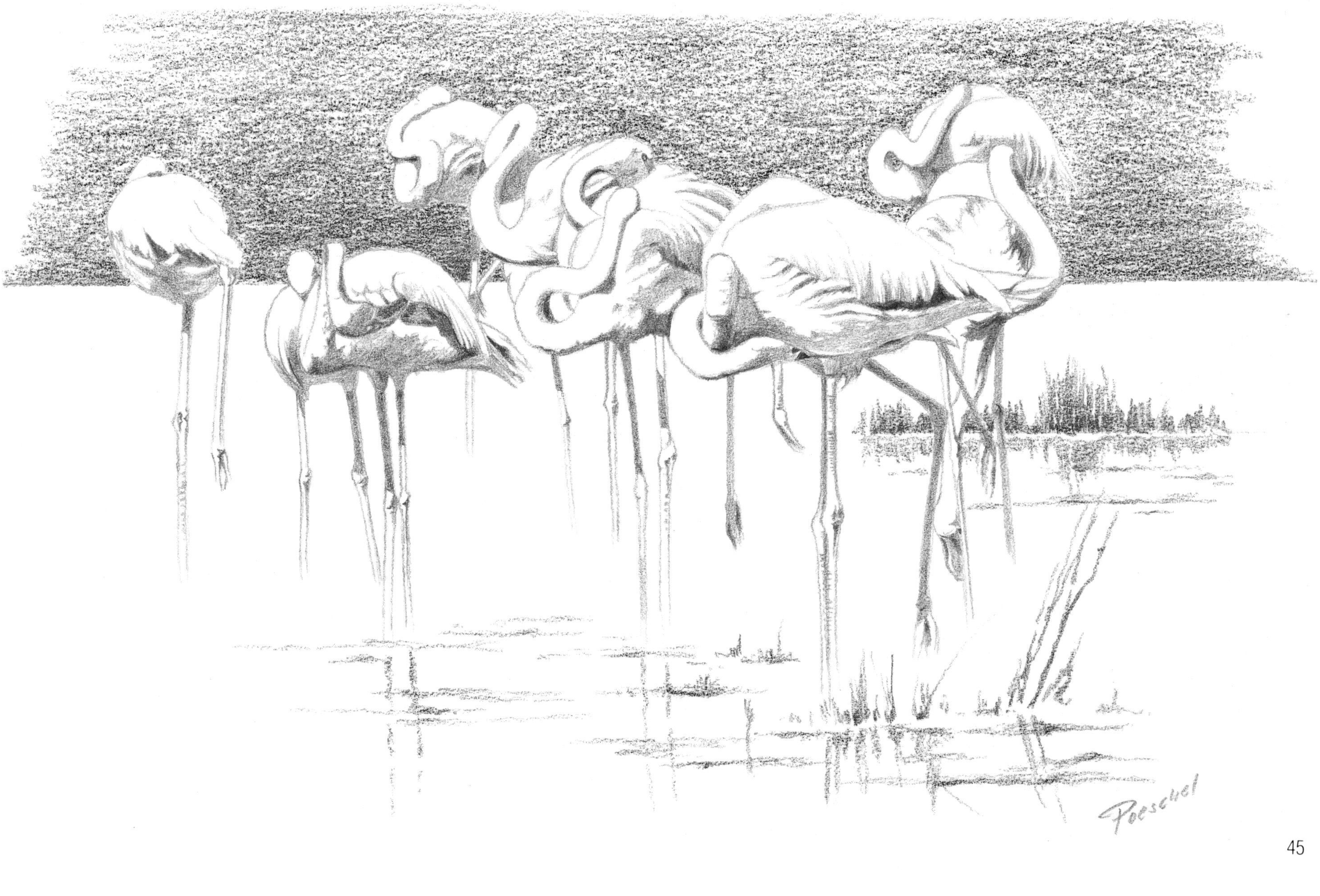
Poeschel

Poeschel

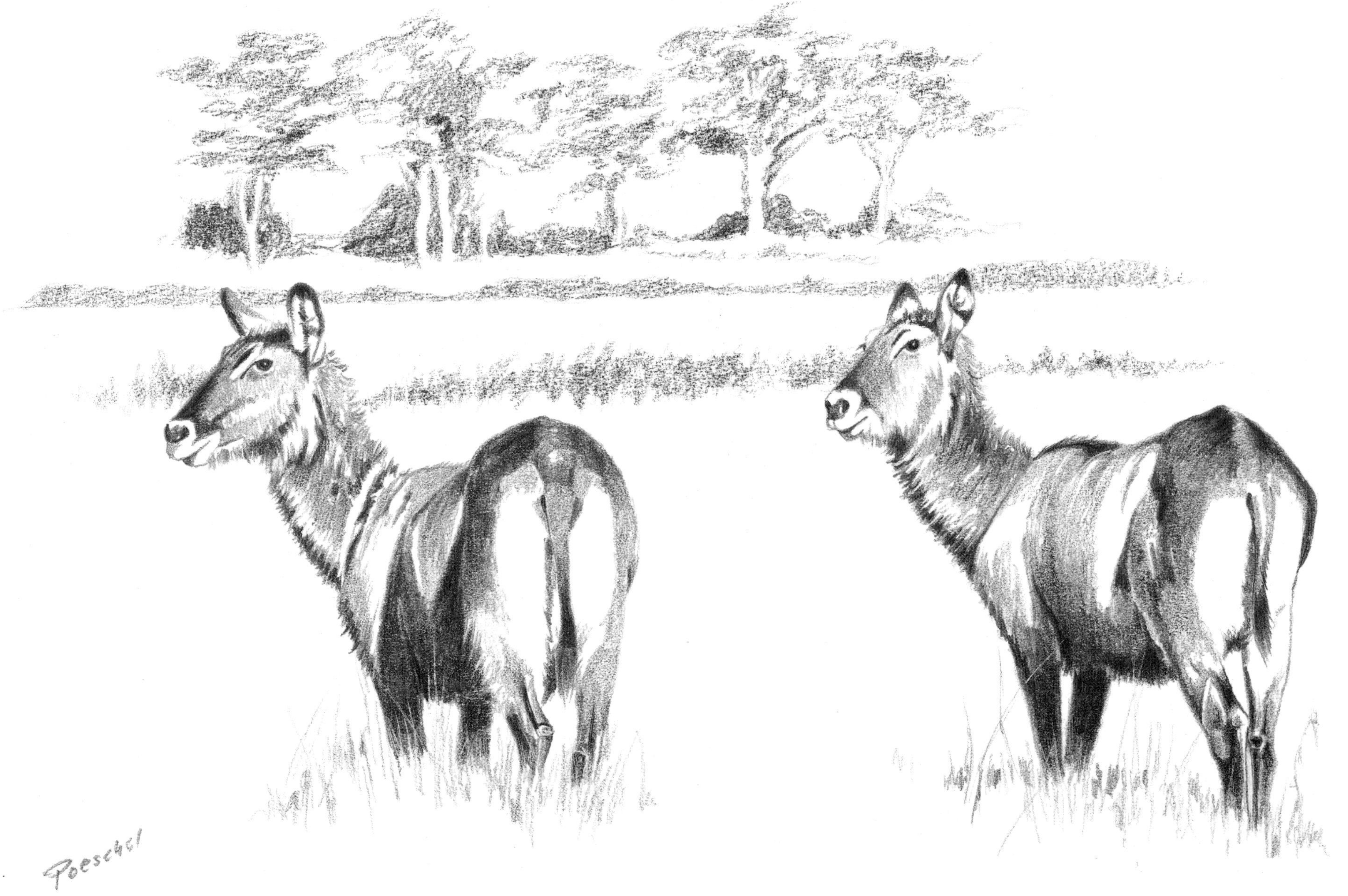
Poeschel

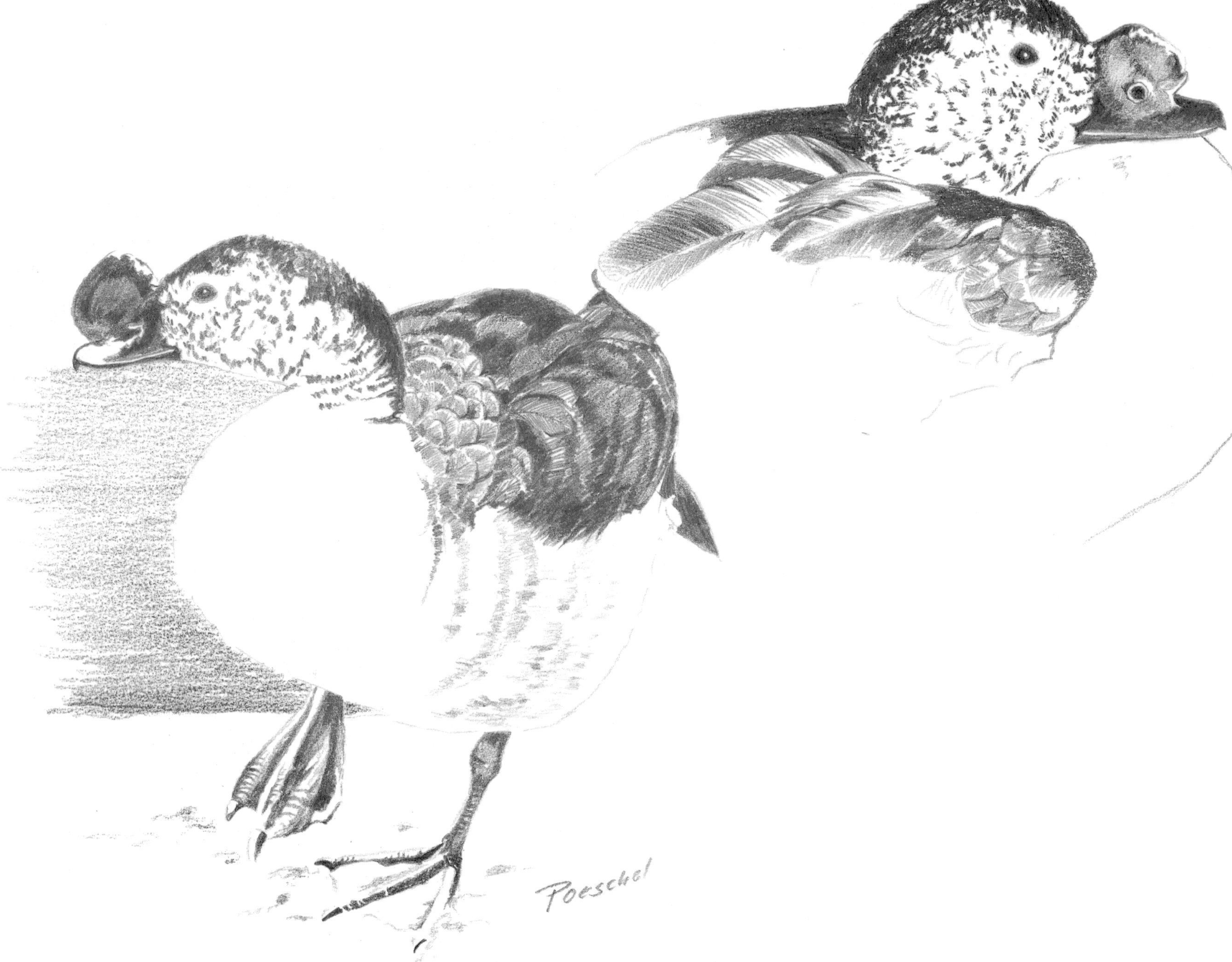
Poeschel

49

Poeschel

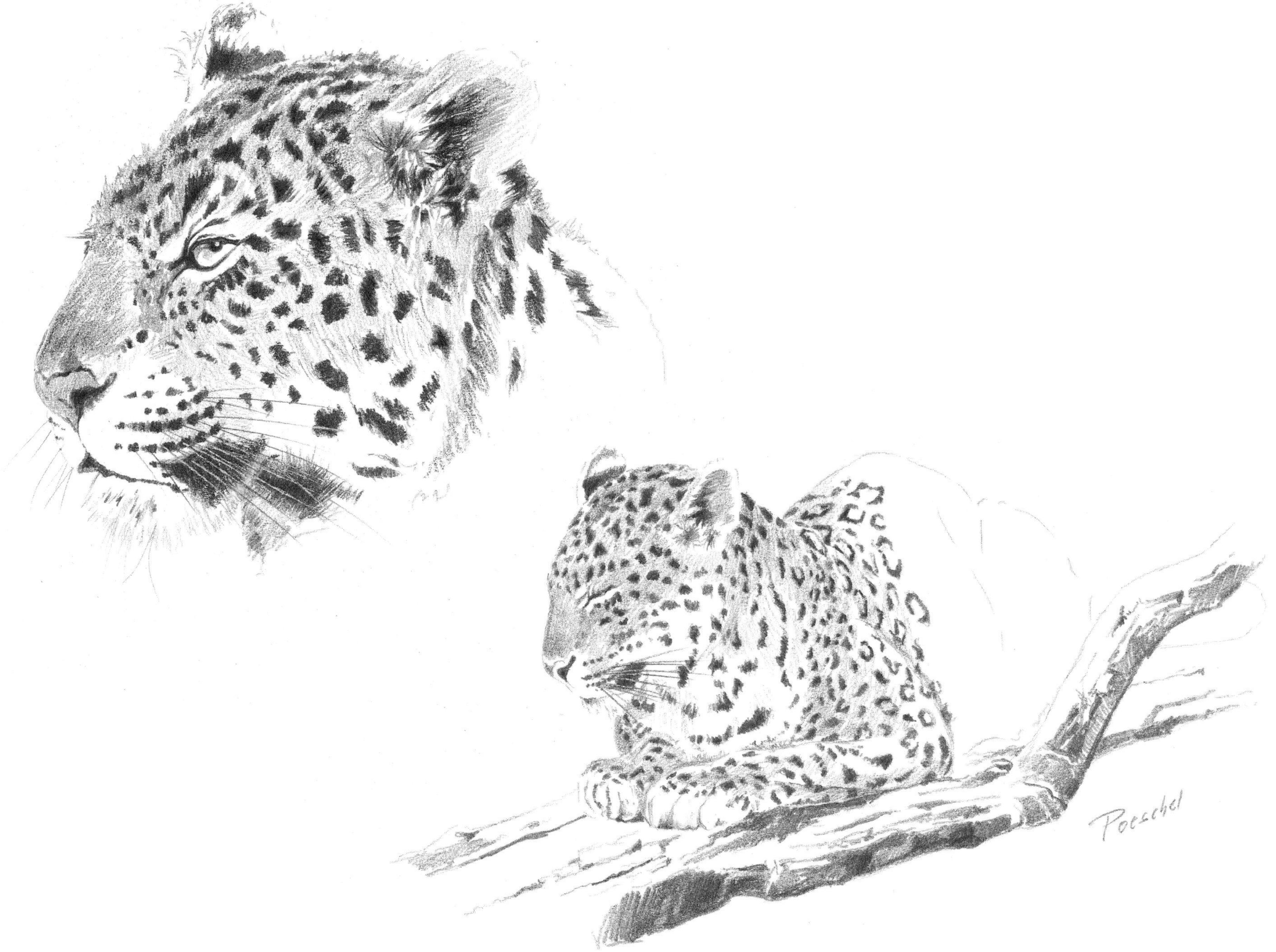
Poeschel

Poeschel

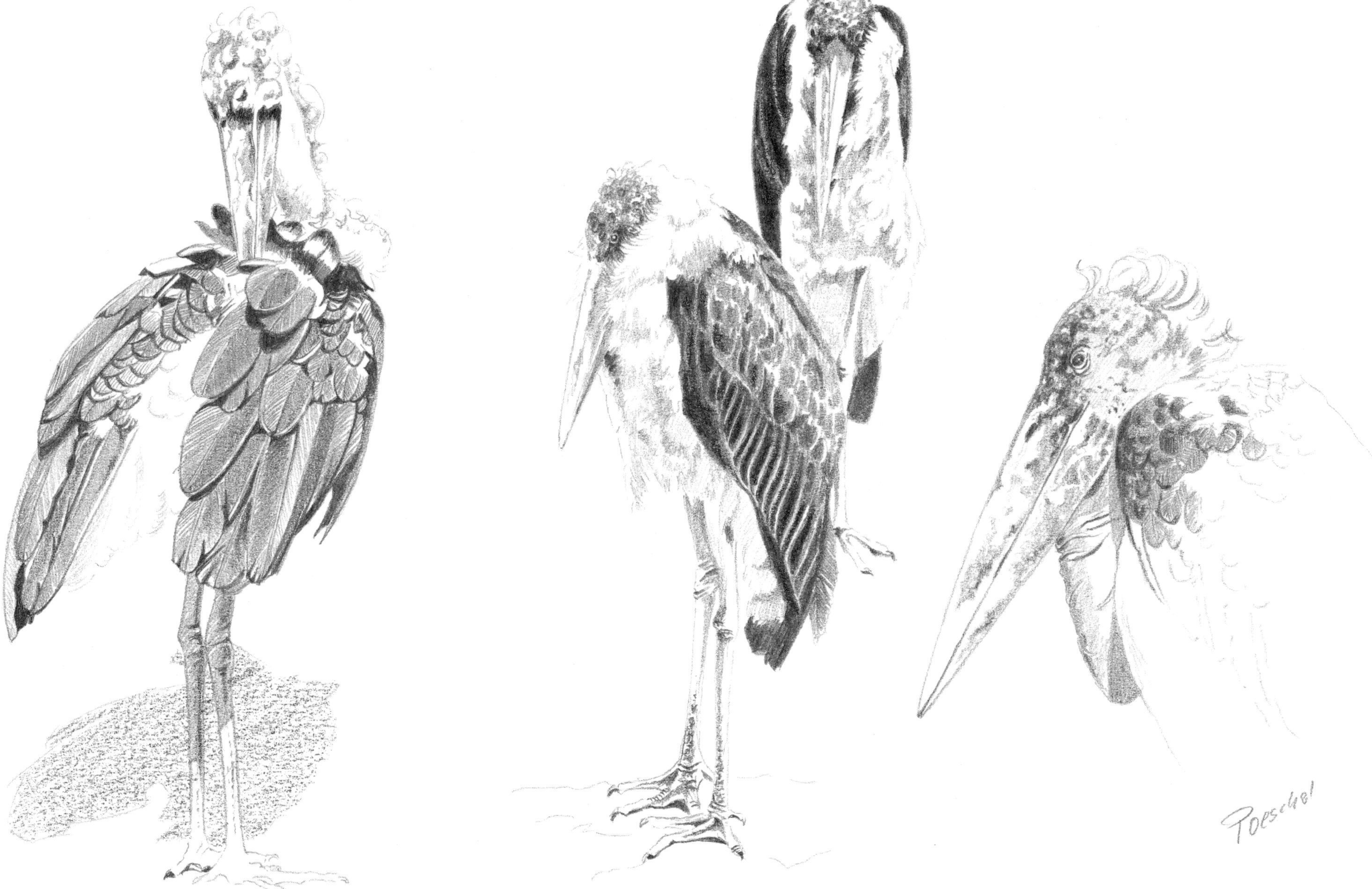
Poeschel

Tanzania

Unter den ostafrikanischen Nationalparks stellt der Manyara ein besonderes Kleinod dar. Am Westrand des Großen Grabenbruchs gelegen ist er mit einer Ausdehnung von 325 qkm einer der kleinsten in Afrika. Früher ein beliebtes Jagdgebiet, wurde er 1957 zum Wildreservat und 1960 zum Nationalpark erklärt.

Unterschiedliche und vielseitige Landschaftsformen sind für den Manyara Nationalpark typisch: Grundwasserwald, vorwiegend mit Schirmakazien bestandene Baumsavannen, Sumpf- und Grasland. Der Marang-Wald am Steilhang des Grabens ist besonders für die Wanderungen der Elefanten von Bedeutung, welche hier in einer Dichte wie in keinem anderen Nationalpark leben. Studien des Forscherehepaares I. und O. Douglas-Hamilton über Sozialverhalten, Gliederung der Familienverbände, Vegetationsschäden usw. wurden in dem Buch »Unter Elefanten« (engl. Titel: »Among the Elephants«) veröffentlicht.

Der Grundwasserwald ist eine Besonderheit. In dem porösen vulkanischen Grund der 50 km entfernten Bergwälder versickert Wasser und tritt in den unteren Lagen des Steilhangs wieder zu Tage. Sykomoren und Tamarinden gedeihen zusammen mit üppigem Bodenbewuchs. Die Savannengebiete sind Lebensbereich der Zebras sowie verschiedener Huftiere wie Impalas, Giraffen und im Sumpfgebiet des Nordteils einer großen Herde von Kaffernbüffel. »Die Attraktion« sind »Baumlöwen«, welche hier einen großen Teil ihrer Ruhepausen verbringen. Nur noch vom Ruwenzori-Nationalpark in Uganda ist dieses Verhalten bekannt. Im Manyara soll der Grund dafür die in Bodennähe lebende Tsetse-Fliege sein.

Verschiedene, je nach Jahreszeit mehr oder weniger Wasser führende Flüsse wie der Ndala, Bagayo und Endabash, münden in den See. Vom Fuß des Grabenrands bis zum Seeufer sind es manchmal nur wenige hundert Meter. Besonders reizvoll ist die Fahrt unmittelbar am Seeufer entlang, wo an Flußmündungen der Süßwasserzuläufe eine besonders reichhaltige Vogelwelt zu sehen ist. In dem dichten Buschgelände wird man oft auf kürzeste Distanz mit Großtieren konfrontiert, im südlichen Waldland »hautnah« mit Elefanten.

Zum Schluß verdient erwähnt zu werden, welch große Anstrengungen afrikanische Länder wie Tanzania unternommen haben, um ihre einmalige Tier- und Pflanzenwelt zu erhalten. Es ist auch unsere Aufgabe, dieses Vorhaben zu unterstützen, um die kostbaren Naturschätze zu erhalten.

Among the East African national parks the Manyara can be considered a special jewel. Situated at the western edge of the Rift Valley, it is with an extension of 325 square kilometers one of the smallest parks in Africa. Formerly a favorite hunting area, it was declared a game reserve in 1957 and a national park in 1960.

A number of different vegetation types are found in the Manyara Park: Forest, savannah with acacias, marshy stretches, and grassland. The Marang forest on the steep slope of the Rift Valley is important for the migrations of the elephants that live here in a denser population than in any other national park. Studies on social behavior, feeding biology and the damage to the vegetation have been published in the book »Among the elephants« by I. and O. Douglas-Hamilton. The groundwater forest is a peculiarity. In the porous volcanic soil water oozes off in the montane forest 50 km away and comes out again in the lower strata of the steep slope. Sycomores and tamarinds flourish here, with a lush undergrowth. The savannah is the habitat of the Zebra and other ungulates as Impala, Giraffe and, in the marshy flat in the north, a large herd of Buffalo. One attraction is the »tree lions« that spend most of their resting time on trees. The only other place where this behavior is known is the Ruwenzori National Park in Uganda. In Manyara it is supposed that the lions are avoiding the tsetse-fly which lives near the ground.

A number of tributaries bring water to Lake Manyara, including the rivers Ndala, Bagayo and Endabash. The edge of the lake is in some places very close to the foot of the steep slope of the Rift Valley. It is especially fascinating to travel along the lake-shore, where on the mouths of rivers and brooks numerous species of birds can be watched. Often in the dense bush big game may be suddenly encountered, something which happens with elephants in the southern woodland.

The great efforts of such African countries as Tanzania to preserve their unique fauna and flora deserve mention. We should consider it our task to support these efforts as much as we can.

Lake Manyara National Park
Tanzania

Poeschel

59

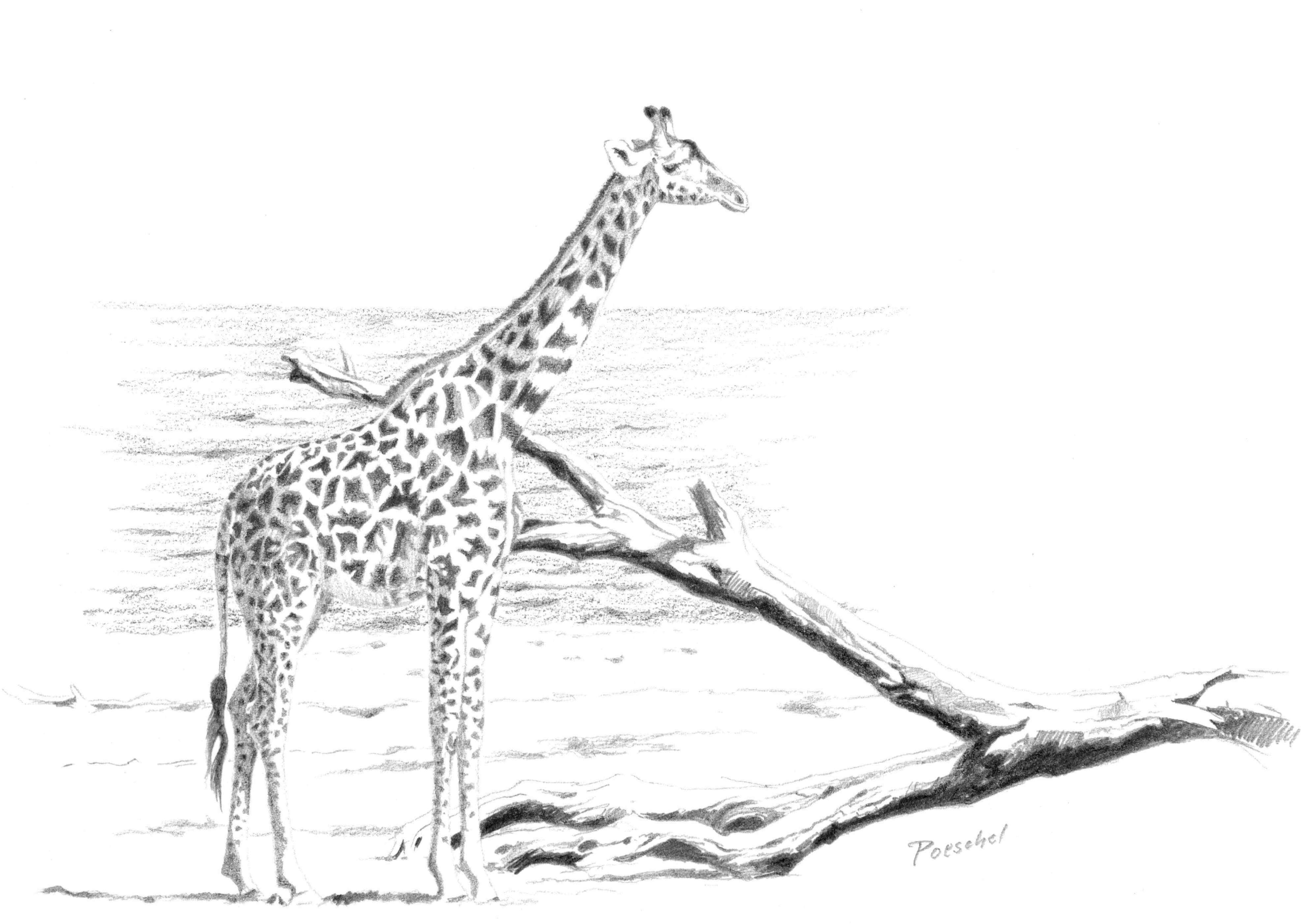
Poeschel

Poeschel

Poeschel

63

Poeschel

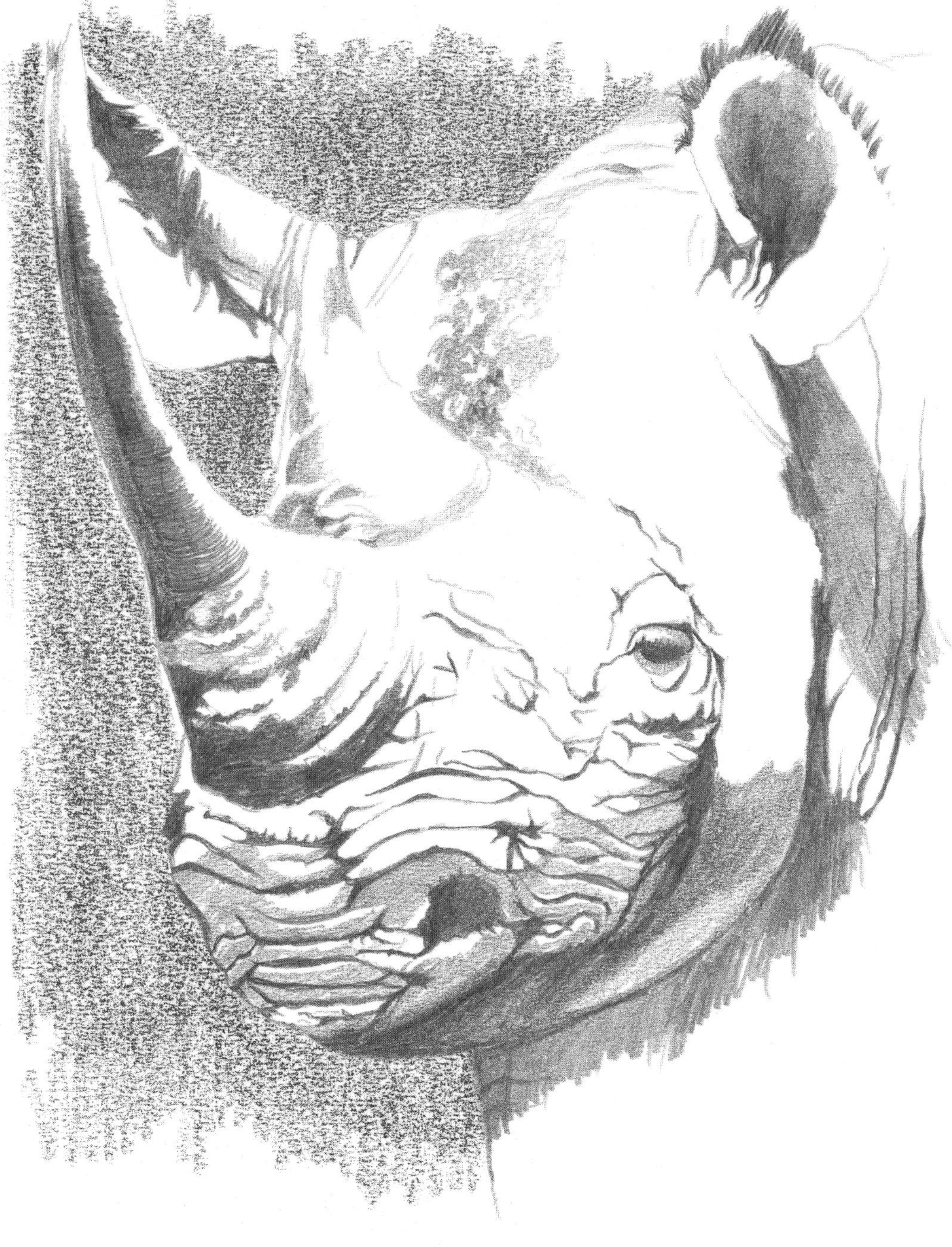

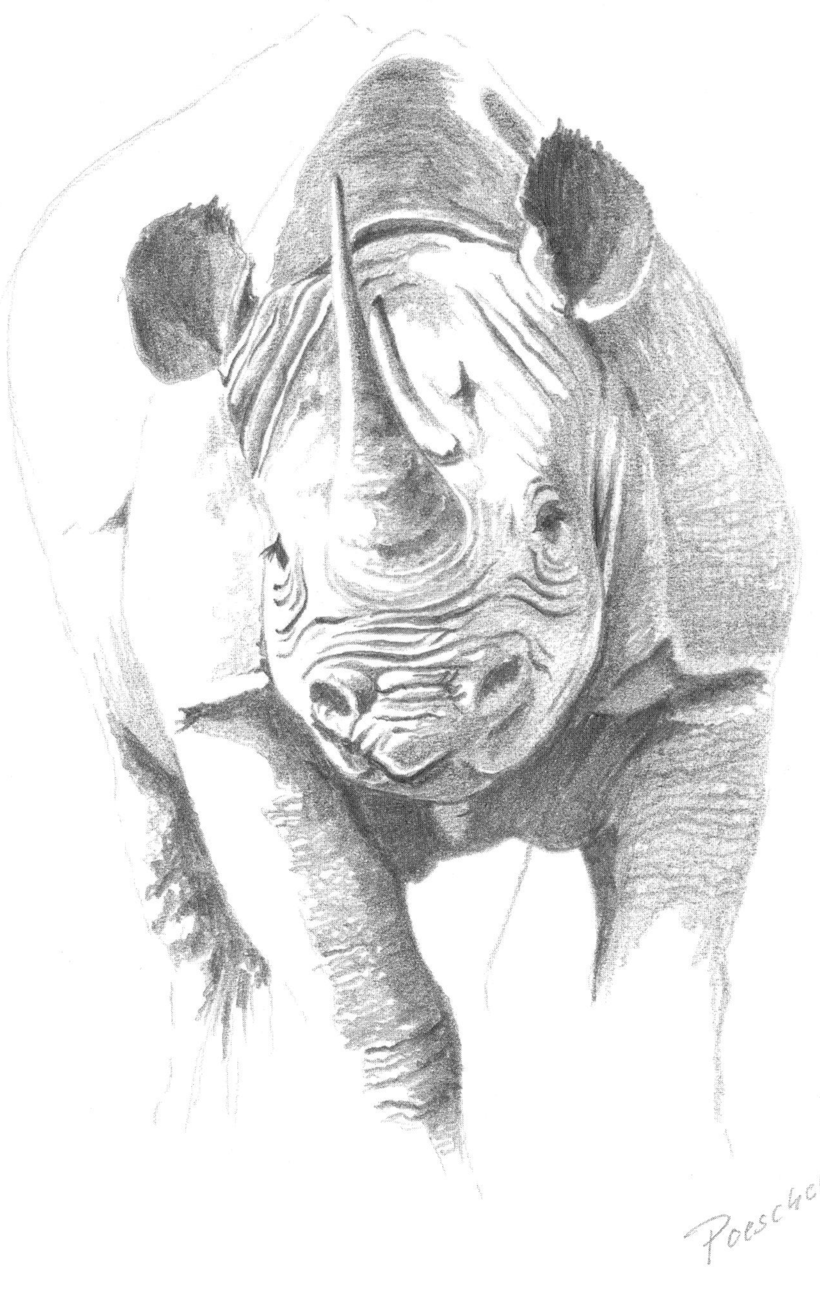
Poeschel

Poeschel

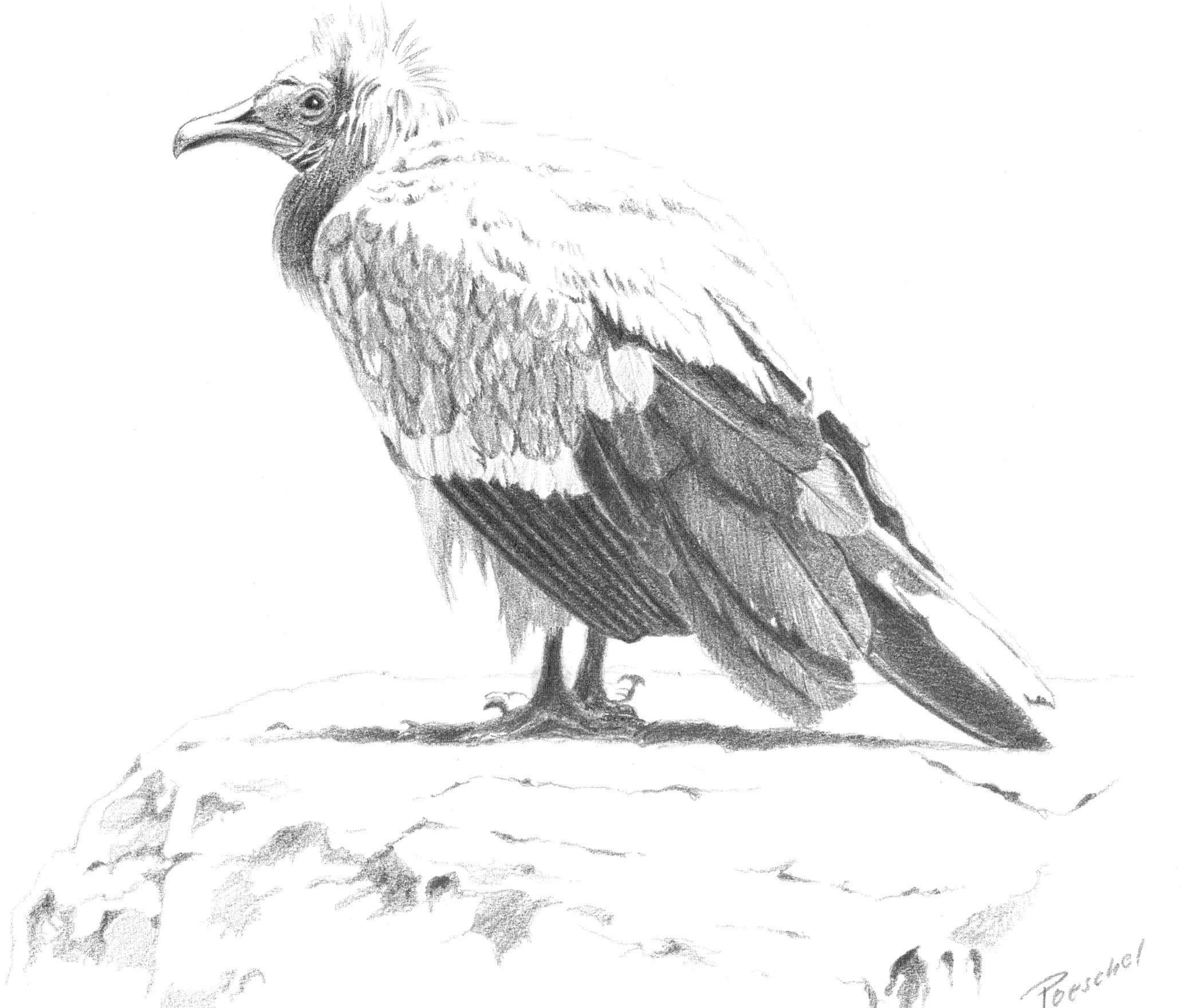
Poeschel

Die Serengeti ist der bekannteste Nationalpark Afrikas. Die Ergebnisse der wissenschaftlichen Arbeit von Prof. Bernhard Grzimek und seinem Sohn Michael, der dabei ums Leben kam, wurden mit dem Buch »Serengeti darf nicht sterben« weltweit bekannt. Der Begriff des Naturschutzgedankens wurde somit an viele Menschen herangetragen und es ist sein Verdienst, daß er mit Nachdruck auf die rapide fortschreitende Zerstörung der Natur aufmerksam gemacht hat.

Die Serengeti ist das letzte Gebiet der Erde, in welchem Wanderungen großer Tierherden, dem Rhythmus des jahreszeitlichen Nahrungsangebotes folgend, ohne Beeinträchtigung durch menschliche Eingriffe noch stattfinden können. 1929 wurde etwa die Hälfte seiner heutigen Fläche zum Schutzgebiet erklärt und im Jahre 1951 zum Nationalpark. Einige Teile wie das Ngorongoro-Schutzgebiet wurden später durch die britische Kolonialregierung leider wieder abgetrennt. Die Serengeti ist keine ebene Grasfläche, sondern gliedert sich in verschiedene Landschaftsstrukturen. Die südliche Kurzgrassteppe geht in der Parkmitte bei Seronera in lichte Baumsavanne über. Im Norden, wie auch im angrenzenden Massai-Mara Game Reserve in Kenya, ist die Landschaft hügelig, Akazienwaldungen wechseln mit großflächigen Langgrassteppen. Besonders auffallend sind die sogenannten »Kopjes«, aus Urgestein bestehende Inselberge, die über das ganze Nationalparkgebiet verstreut sind.

Der Anblick der riesigen Tierbestände ist beeindruckend, denn soweit das Auge reicht, erfaßt es kaum die unübersehbaren Herden von Gnus, dazwischen eingestreut Gruppen von Zebras, Topis und Gazellen. Dies vermittelt einen Begriff von dem ungeheuren Tierreichtum, den einmal der afrikanische Kontinent aufzuweisen hatte.

Für mich waren es unvergeßliche Eindrücke und ich möchte an dieser Stelle all jenen danken, die sich dafür eingesetzt und darum gekämpft haben, daß wir dieses grandiose Schauspiel noch erleben dürfen.

The Serengeti is the most famous national park of Africa. The results of the work of Professor Bernhard Grzimek and his son Michael, who lost his life during these activities, became known all over the world with the book »Serengeti must not die«. It is to his merit to have made the public aware of the rapid destruction of nature and to have brought the notion of nature preservation to many people.

The Serengeti is the only area left on earth where the migrations of vast game herds, determined by the rhythm of seasonal food supply, can still take place without being impaired by human activities. About half of today's area was declared a reserve in 1929, and a national park in 1951. Unfortunately some parts, such as the Ngorongoro reserve were excluded later on by the British colonial government.

The Serengeti is not a uniform grassland plain, but a diversified landscape. The southern short grass steppe changes in the centre of the park to a tree savannah. In the north, as in the adjacent Massai Mara Game Reserve in Kenya, the terrain is hilly, acacia woods alternate with larger areas of long grass steppe. Especially conspicuous are the »kopjes«, isolated granite hills found all over the Serengeti Park.

The sight of the huge game herds is very impressive; as far as the eye reaches there are enormous herds of Gnu, interspersed with groups of Zebra, Topi and Gazelle. This gives an idea how rich in animal life Africa once was.
For me these were unforgettable scenes. I wish to thank all those to whose efforts it is due that today we can still experience this great spectacle.

Serengeti National Park
Tanzania

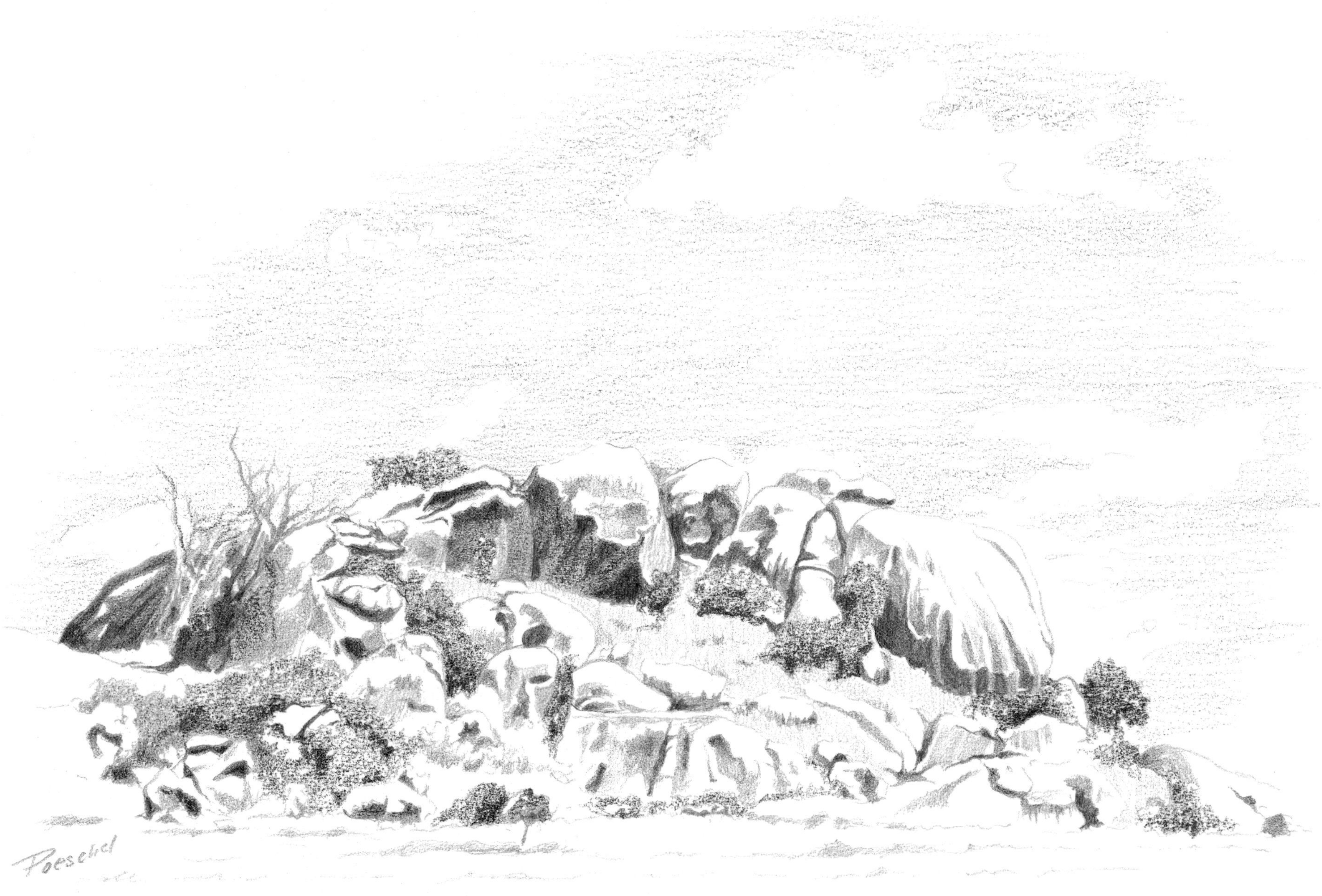
Poeschel

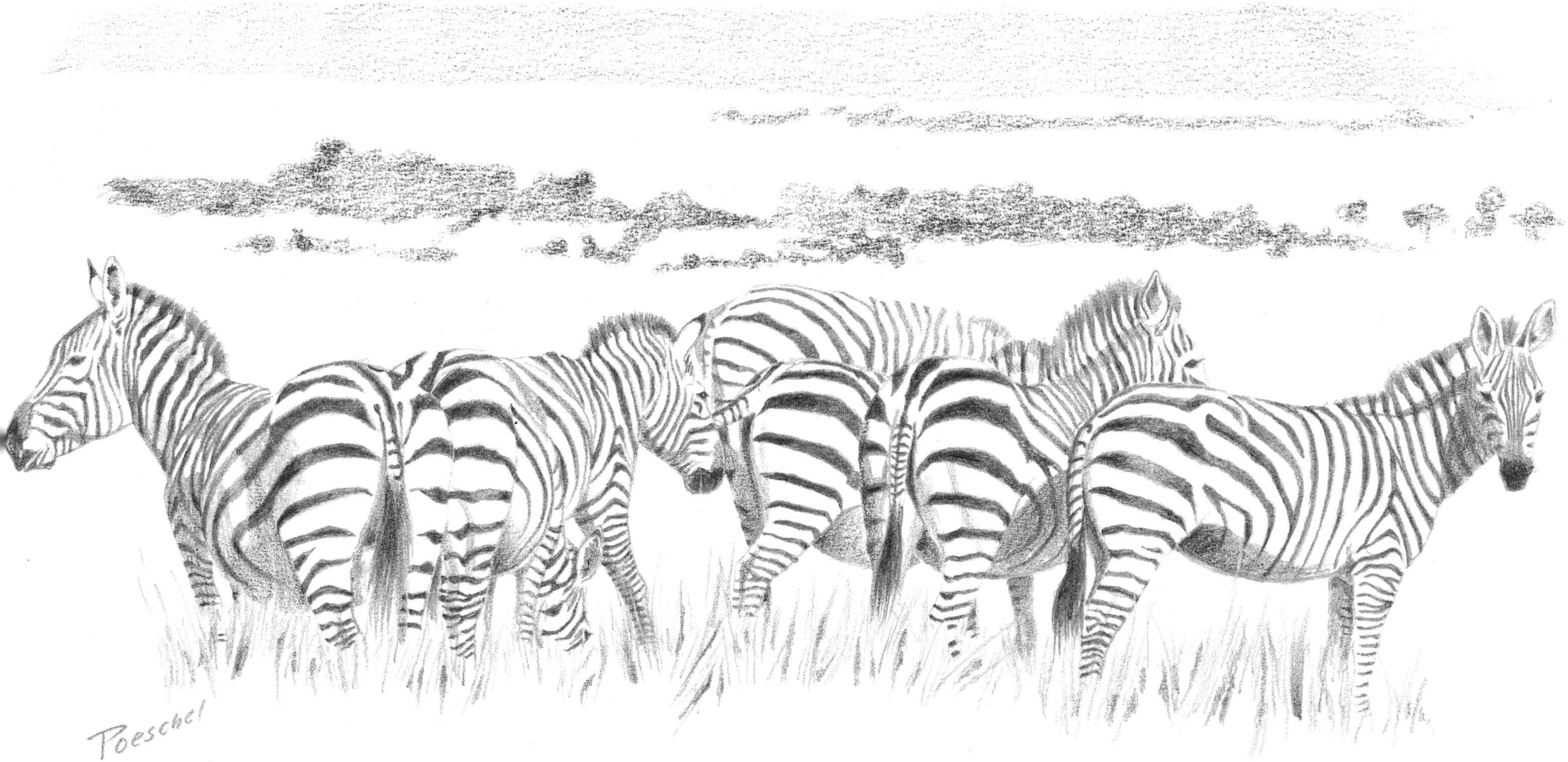
Poeschel

Poeschel

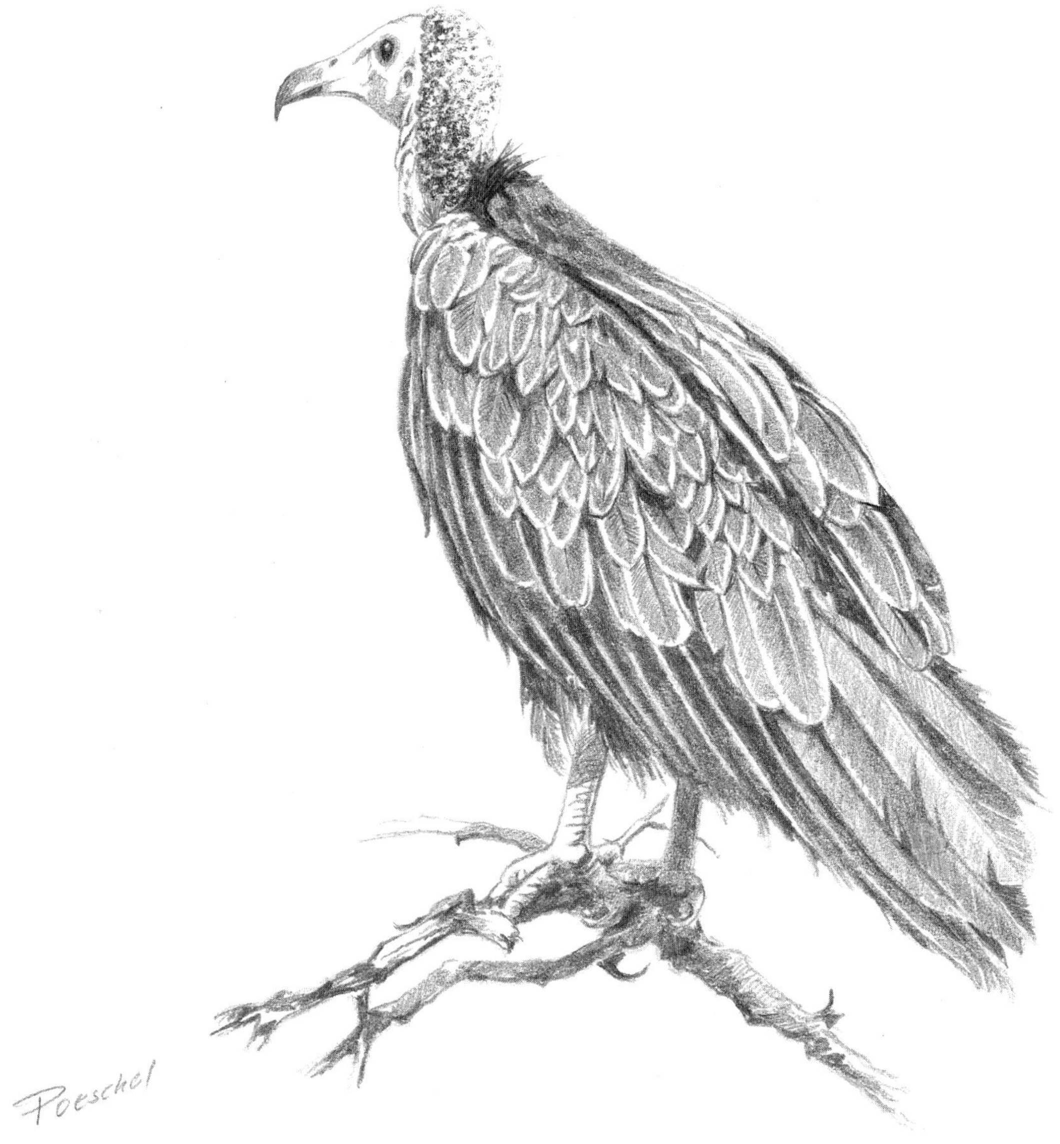
Poeschel

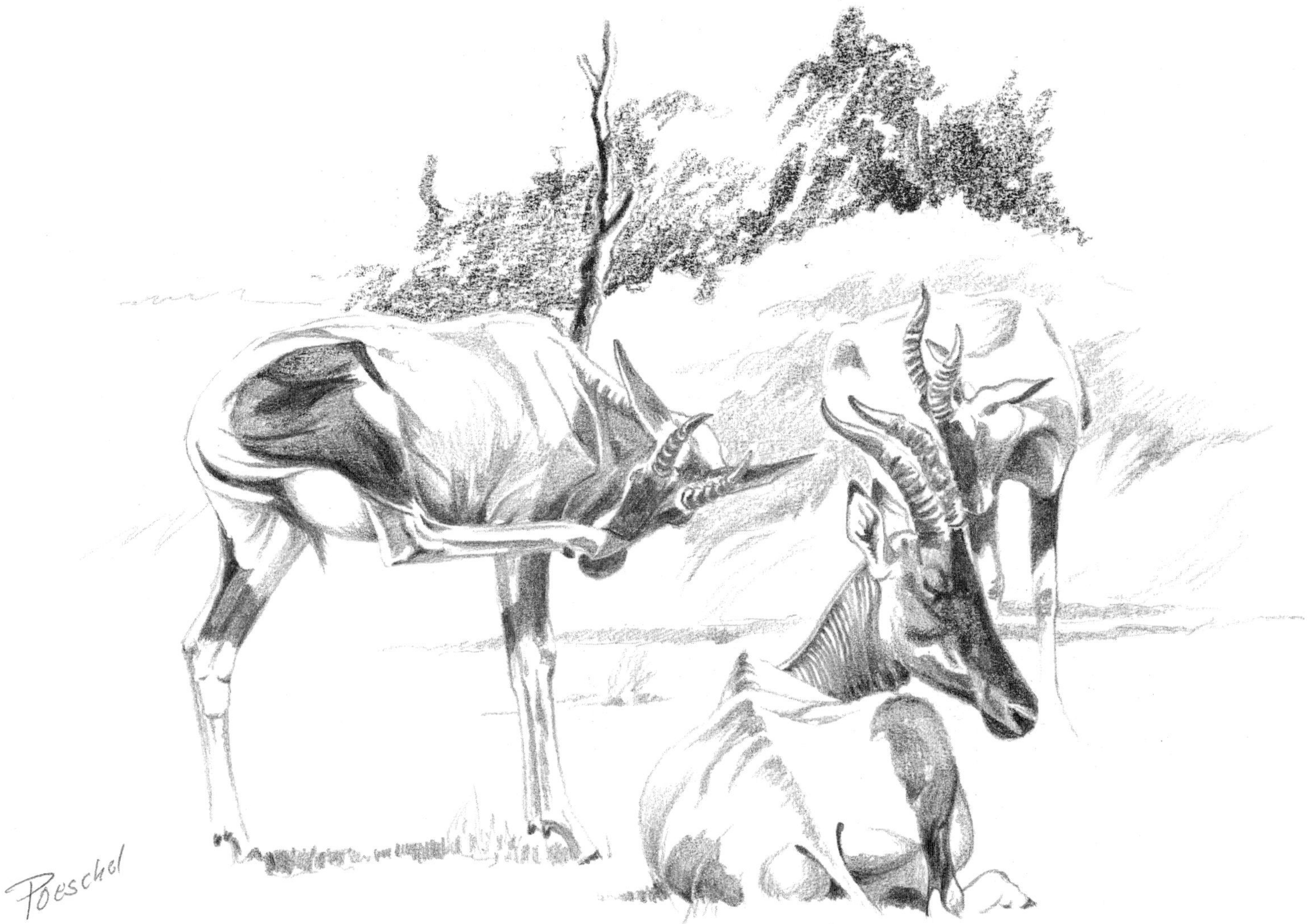
Poeschel

♂

♀

Poeschel

75

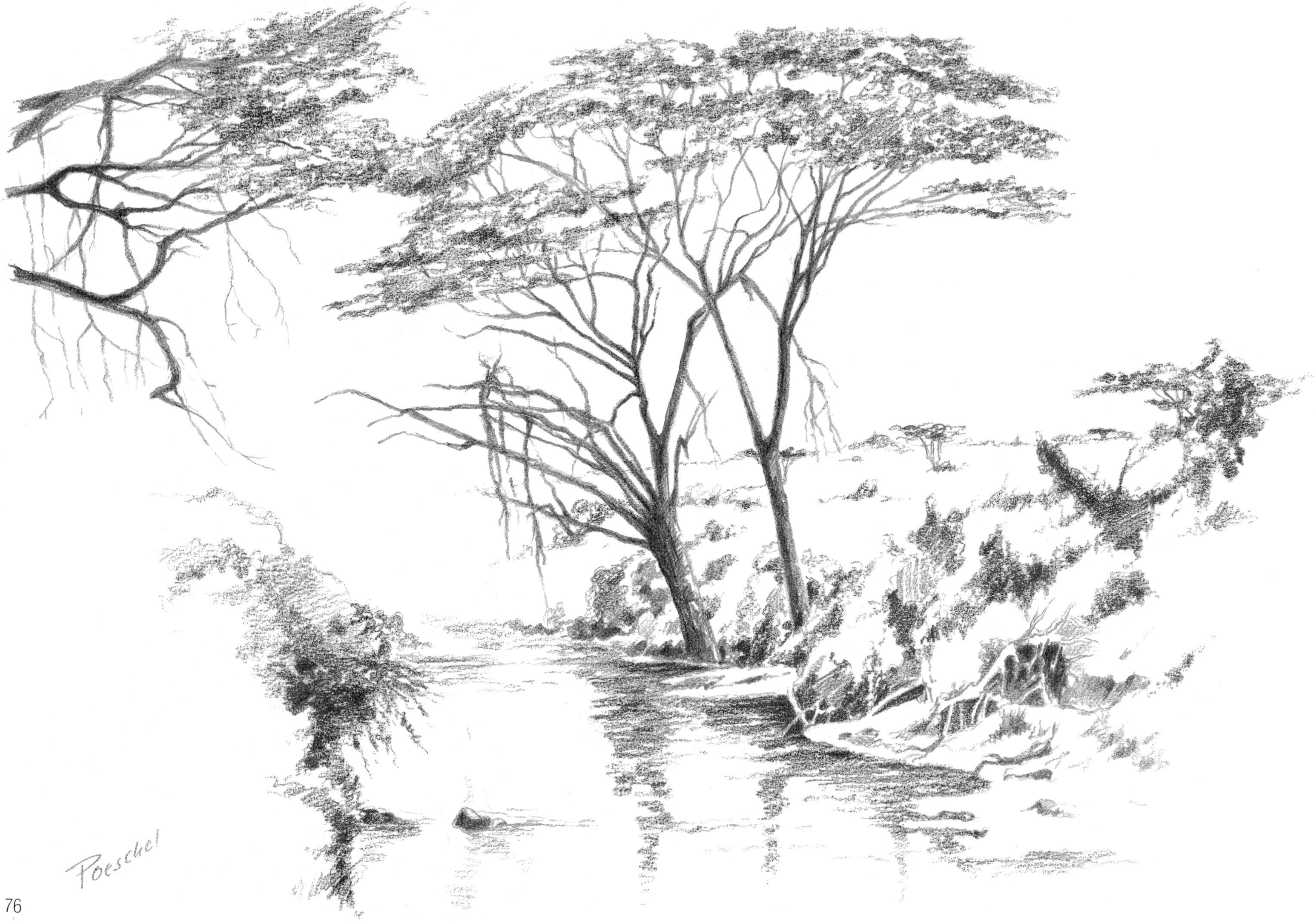
Poeschel

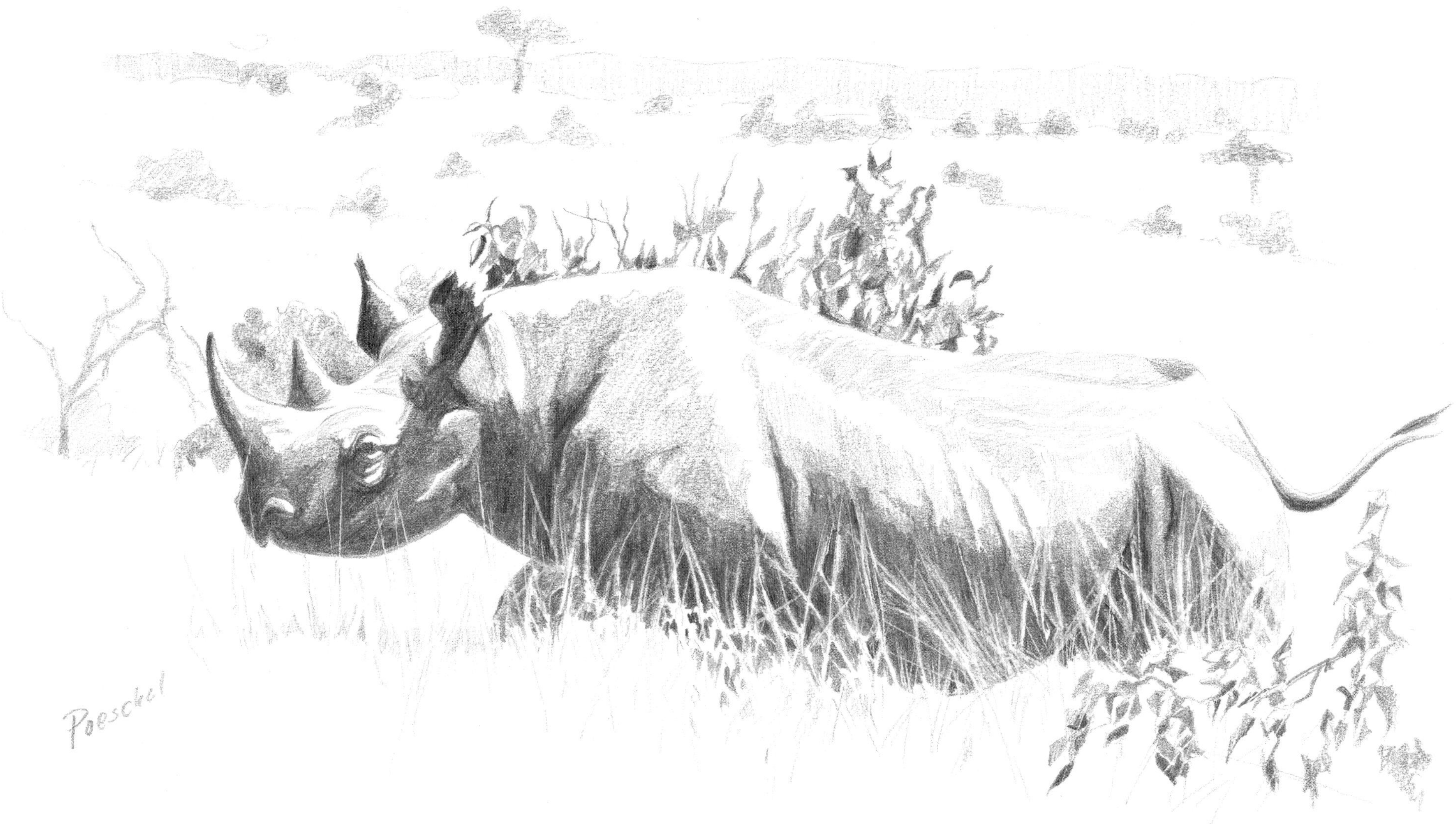
Poeschel

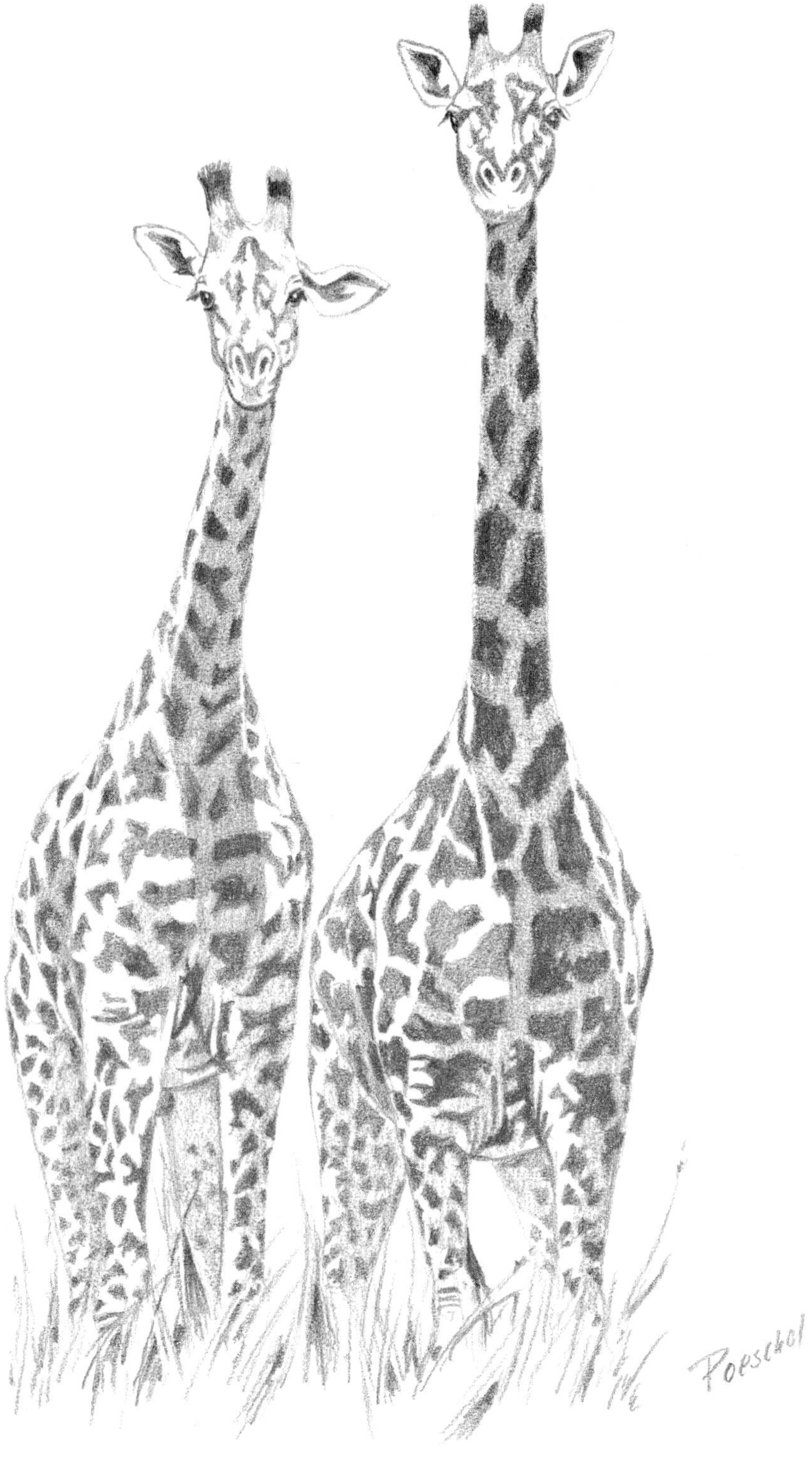
Poeschel

Poeschel

Poeschel

Botswana

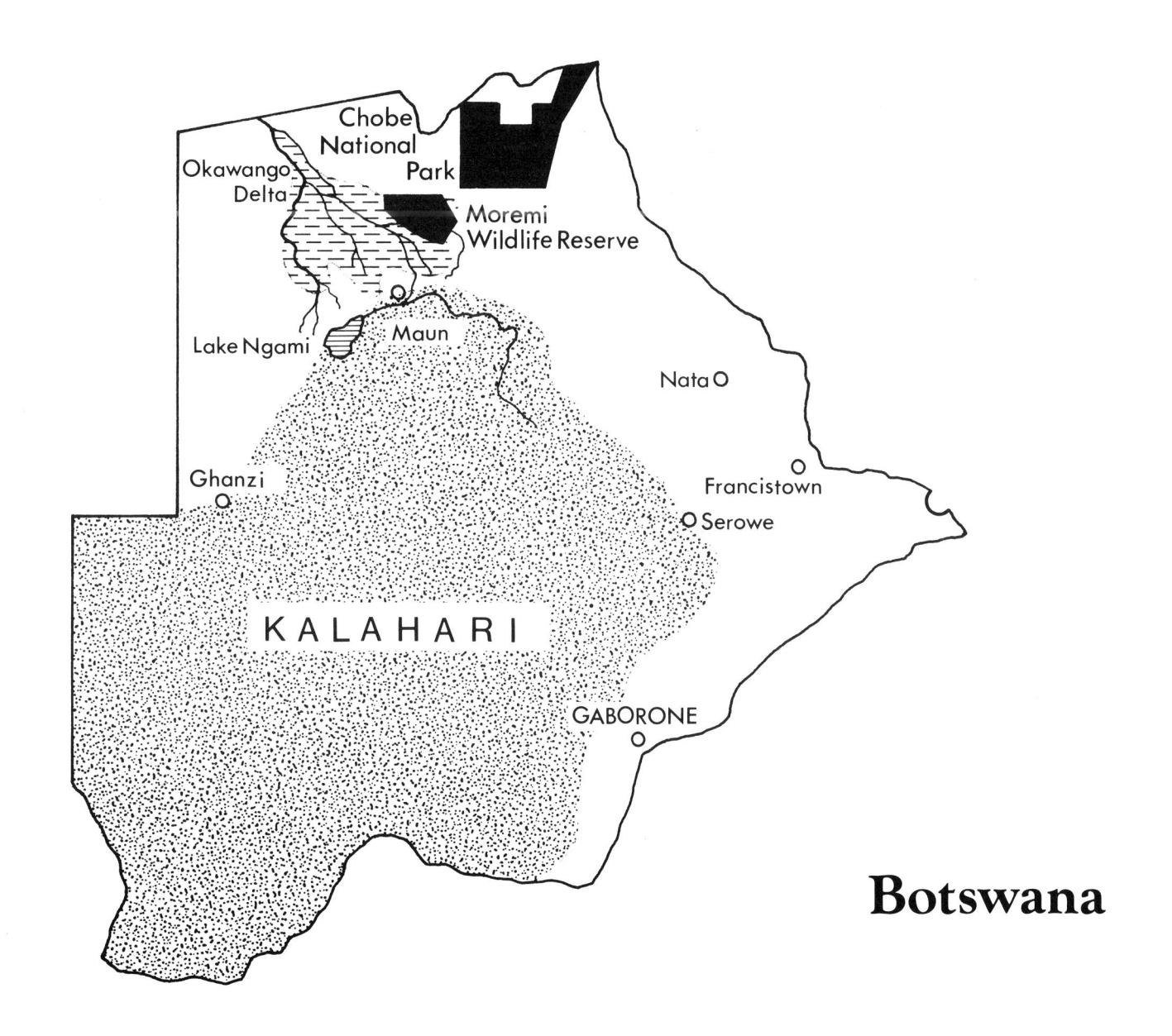

Der Chobe ist der nördliche Grenzfluß Botswanas. Er ist die Weiterführung des Linyanti, der aus dem Okawango-Delta kommt und der an der Vierländerecke, den Staatsgrenzen von Botswana, Südwestafrika (Namibia), Zambia und Zimbabwe, in den Sambesi mündet. Der zu Beginn der 60er Jahre gegründete, 11700 qkm große Nationalpark ist nach dem Chobe benannt. Unmittelbar an seinem Ufer beginnen dichte Busch- und Akazienwälder, welche den Charakter der nördlichen Vegetation bestimmen. Dieses Gebiet ist wegen seiner Elefantenherden bekannt, die bis zu 500 Tiere zählen können.

Am südwestlichen Teil schließt sich die Mababe-Senke (Mababe Depression) an. Hier herrscht die Trockenheit der Kalahari, deren schütterer Steppenvegetation sich Tiere wie Oryx-Antilopen, Springböcke und Strauße angepaßt haben.

Das eigentliche Kerngebiet des Chobe-Nationalparks ist das Gebiet am Savuti-Fluß, der durch eine Gabelung am Linyanti-Dreieck entsteht. 1853 führte der Savuti noch Wasser, danach trocknete er für 100 Jahre völlig aus. In seinem Flußbett wuchsen Bäume. 1957 kehrte das Wasser zurück. Am Ende seines Laufes versickert der Fluß im Savuti-Marsh, einer baumlosen, mit Ried und Elefantengras bewachsenen, topfebenen Fläche. Führt der Savuti Wasser, wird er zum Hauptziel wandernder Tierherden. So weit das Auge reicht sieht man Gruppen von Sassabys, Gnus, Impalas, Giraffen und Zebras und als Besonderheit Rappen- und Pferdeantilopen. Große Büffelherden und eine reichhaltige Vogelwelt bevölkern das Schwemmland. Bei meinem zweiten Besuch 1983 war der Fluß ausgetrocknet, verschwunden waren die Flußpferdherden und die an das Wasser gebundene Tierwelt. Apathisch gruben Elefanten im Flußbett nach Wasser. Die seit Jahren Südafrika heimsuchende Trockenheit hatte auch die Wasser des Savuti zum Verschwinden gebracht.

Savuti! Für jeden Kenner ein Zauberwort. Hier gibt es keinen Massentourismus, hier kann man nicht mit dem Whiskyglas in der Hand auf den Veranden bequemer Lodges die Tierwelt an den Wasserstellen beobachten. Hier ziehen Elefanten durch das Camp. Bei Nacht ist das Grollen der Löwen und das »hu-up« der Hyänen in unmittelbarer Nähe zu vernehmen. Dies ist noch ein Stück echter afrikanischer Natur und für den, der Strapazen auf sich nimmt, ein unvergeßliches Erlebnis.

The Chobe river forms the border of Botswana in the north. It is the continuation of the Linyanti, which originates in the Okawango delta and flows into the Sambesi where the borders of the four countries Botswana, South West Africa (Namibia), Zambia and Zimbabwe touch each other. The national park, named after the Chobe river, covers 11,700 square kilometers and was founded at the beginning of the 1960s. Dense bush and acacia forest border the shores of the river, characterizing the vegetation of the northern part of the park. This region is famous because of its elephant herds, which number up to 500 animals.

The southwestern part is formed by the Mababe Depression, under the influence of the arid climate of the Kalahari desert, with a meagre steppe vegetation and animals adapted to this habitat such as Oryx Antelope, Springbuck and Ostrich.

The centre of the Chobe National Park is the area of the Savuti river. This river flowed until 1853, afterwards it was completely dry for 100 years and trees grew in its bed. Since 1957 water has been running in the river again, at least seasonally. The course of the river ends in the Savuti Marsh, a treeless plain overgrown with reed and elephant grass. At times when the river has water it attracts the migrating game herds of the surrounding region. Along its shores it is teeming with grazing parties of Sassaby, Gnu, Impala, Giraffe and Zebra, and occasionally the rare Roan Antelope and Sable Antelope. Large herds of Buffalo and a rich bird life populate the flooded areas. During my second visit in 1983 the river was dry, and the herds of hippo and other water animals had disappeared. Elephants dug for water in the dry river bed. The droughts haunting South Africa for years had now also caused the water of the Savuti river to disappear.

Savuti is a magic word for every connoisseur. There is no mass tourism, and it is not possible to watch the game at the watering place at close quarters with the whiskey glass in hand, lounging on the verandah of a comfortable lodge. Here the elephants trod right through the camp. In the night the growling of the lions can be heard nearby, accompanied by the »hoo-oop« of the hyaenas. This is still a piece of original African nature and reward for those who cope with the exertions required for a visit to this place.

Chobe National Park
Botswana

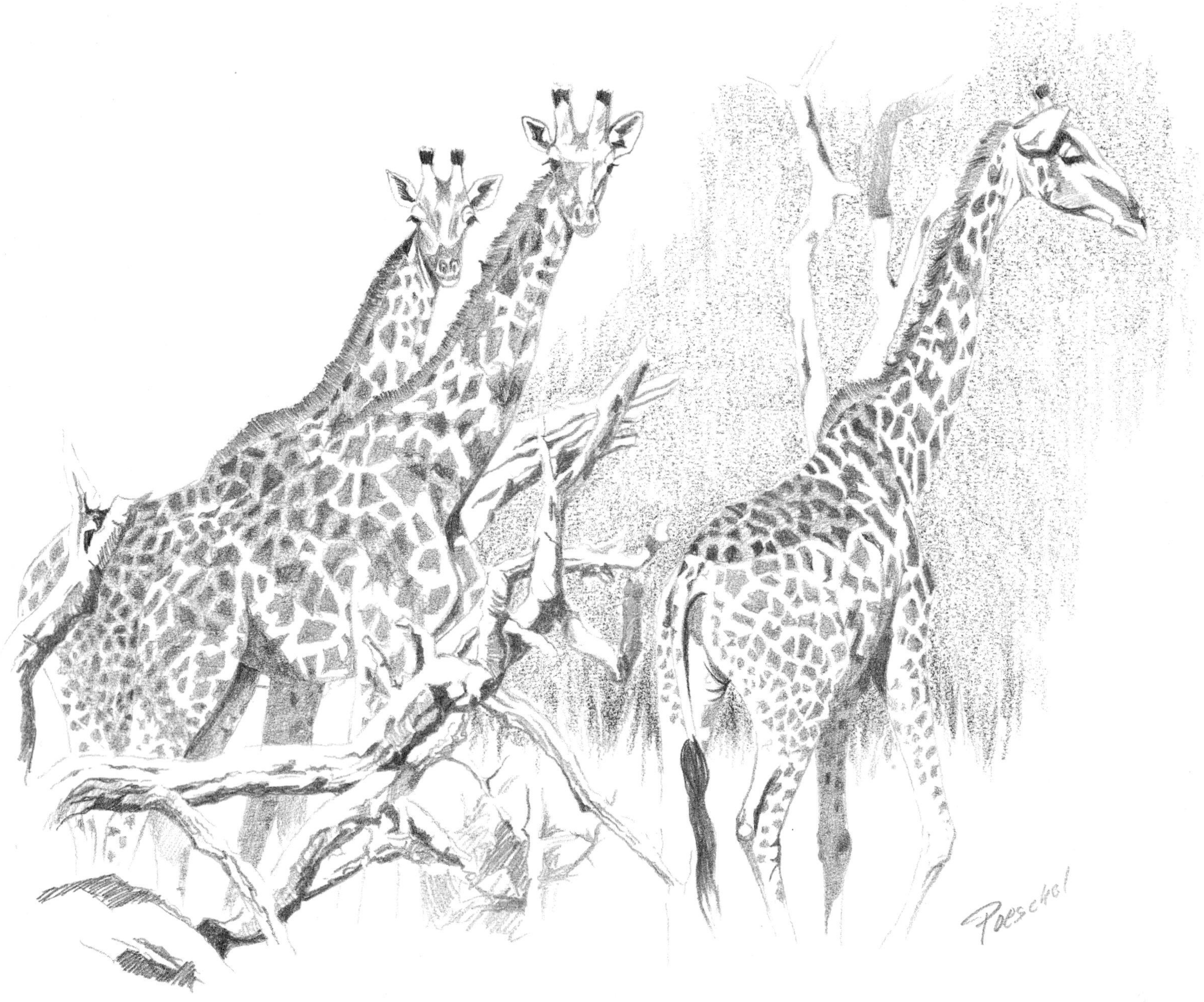
Poeschel

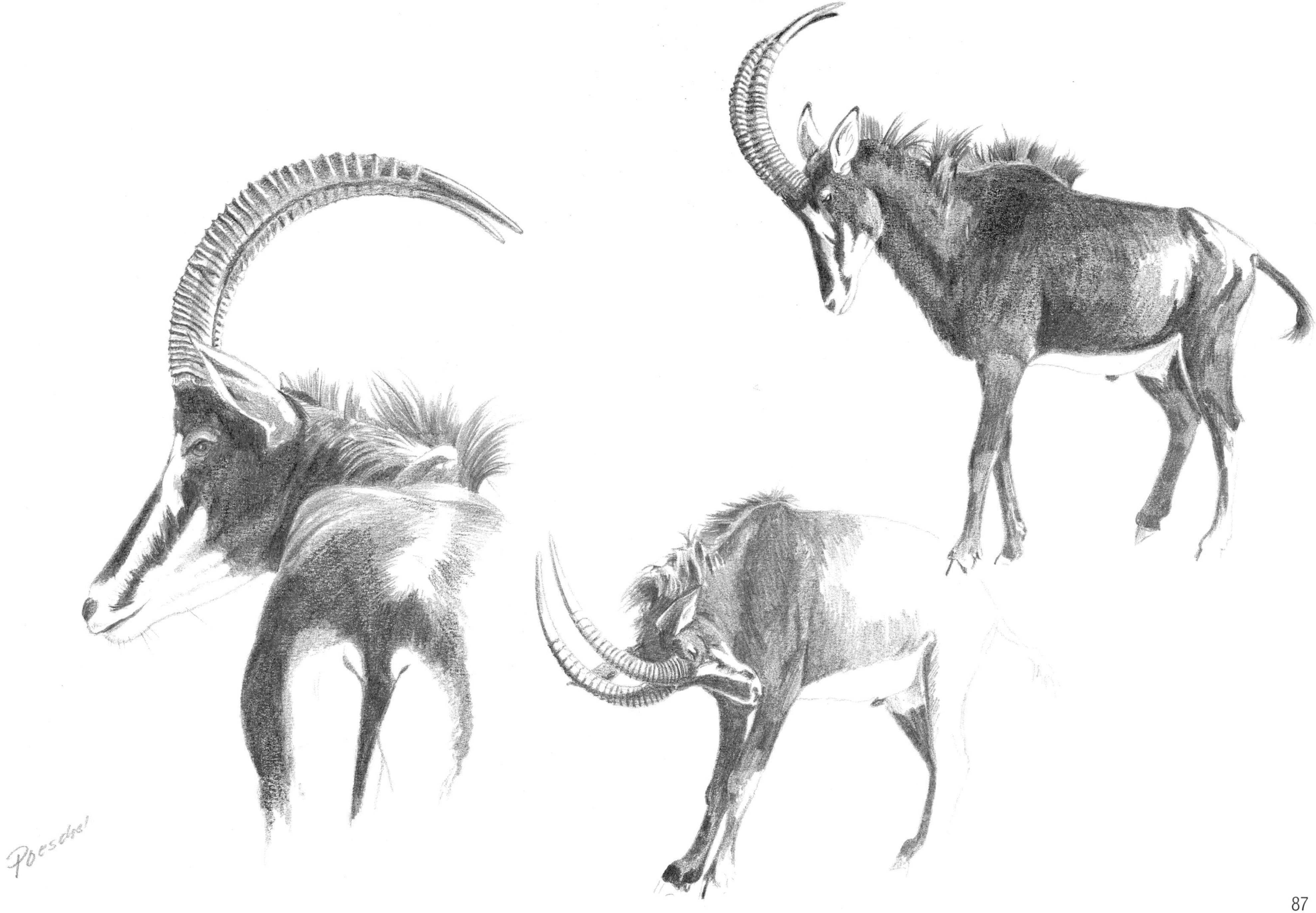
Poeschel

Poeschel

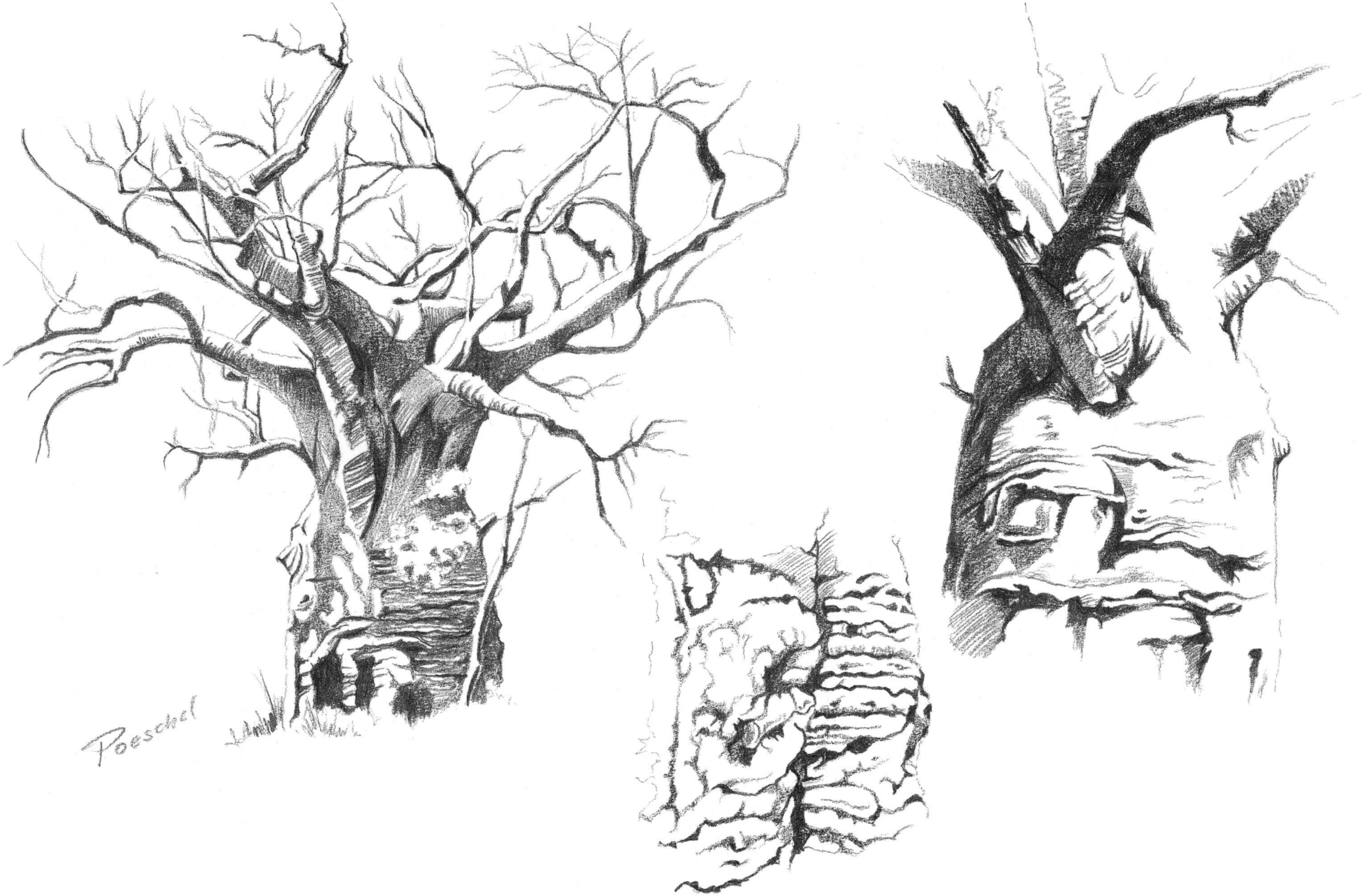
Poeschel

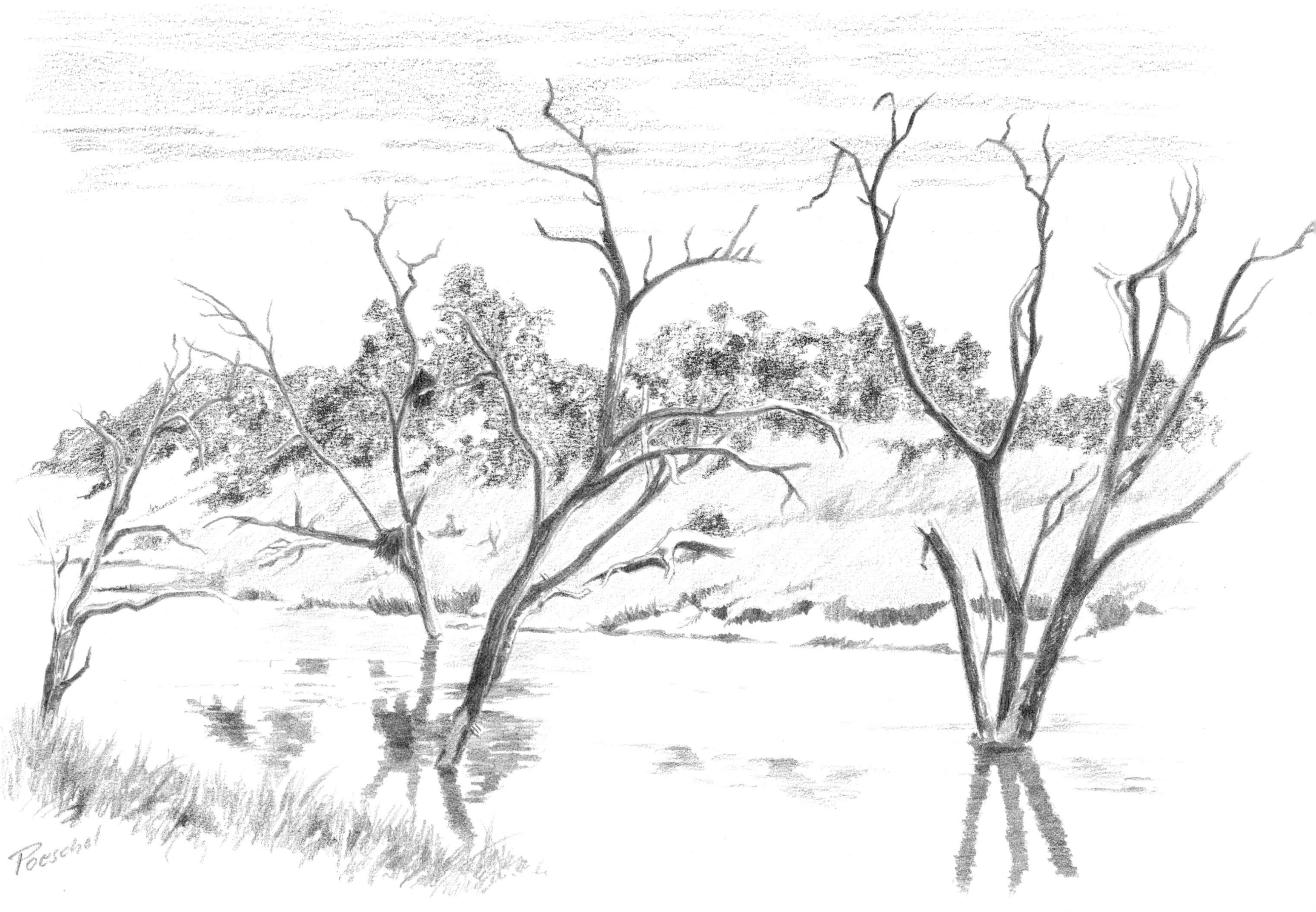
Poeschel

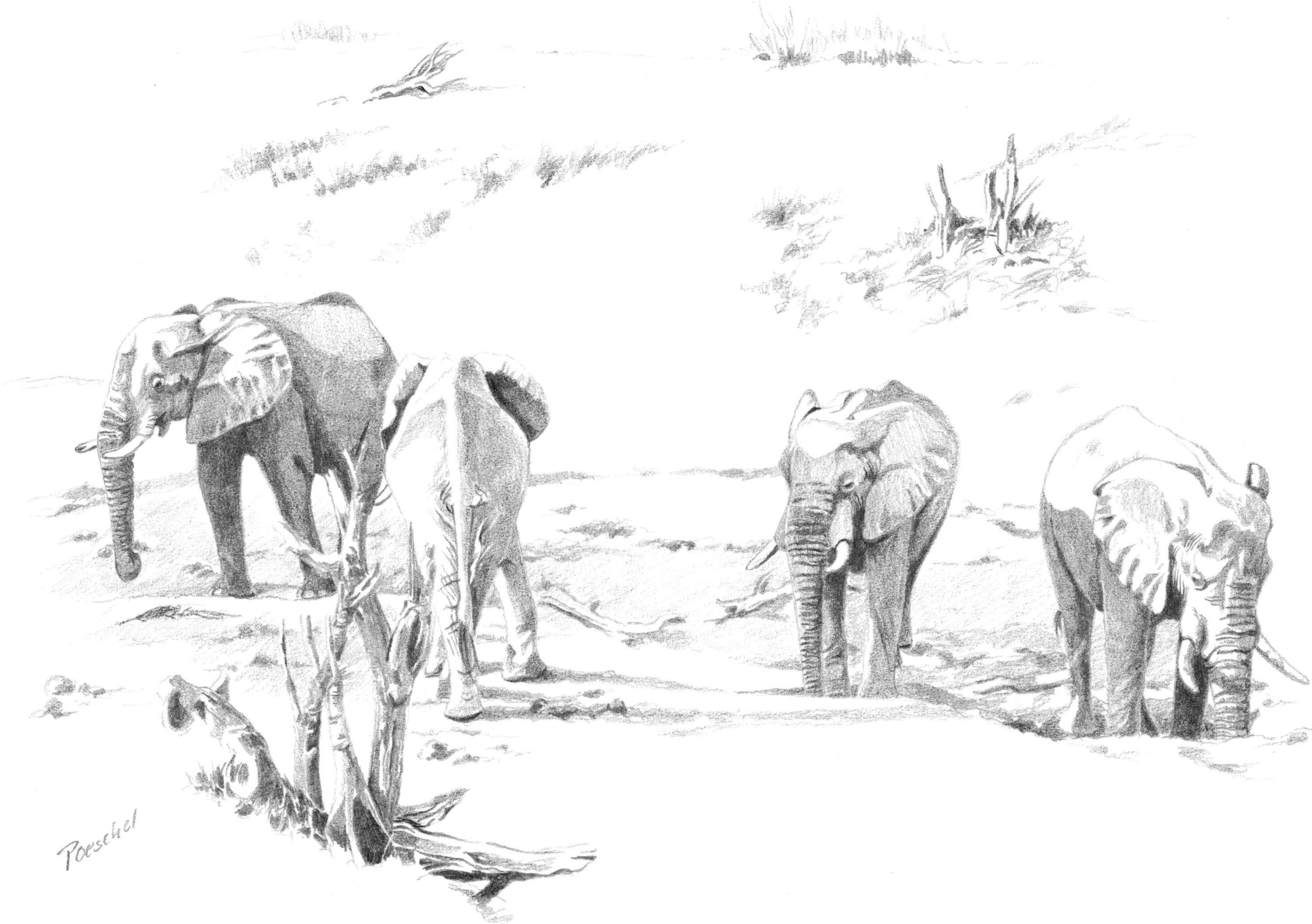
Poeschel

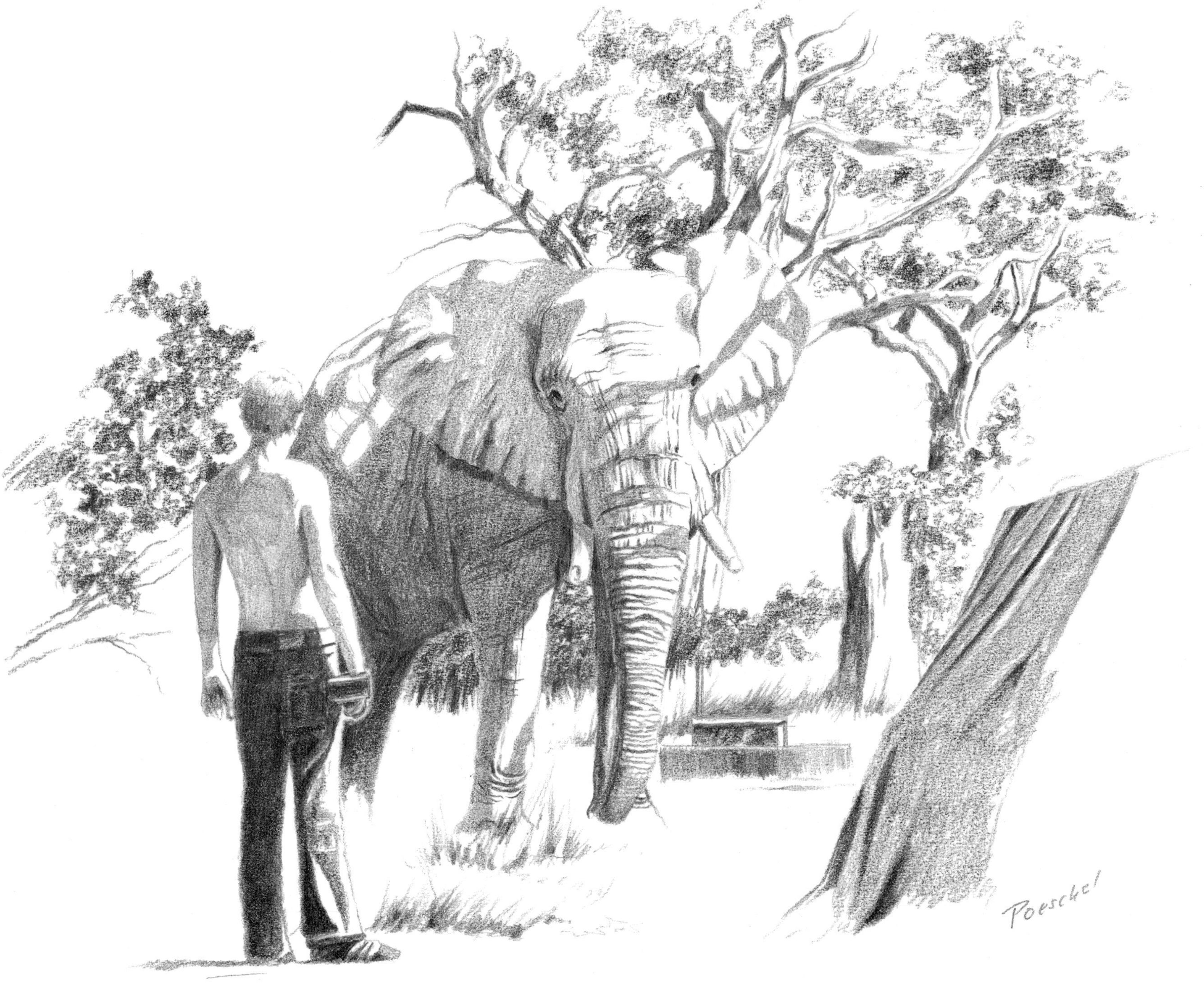
Poeschel

Poeschel

94

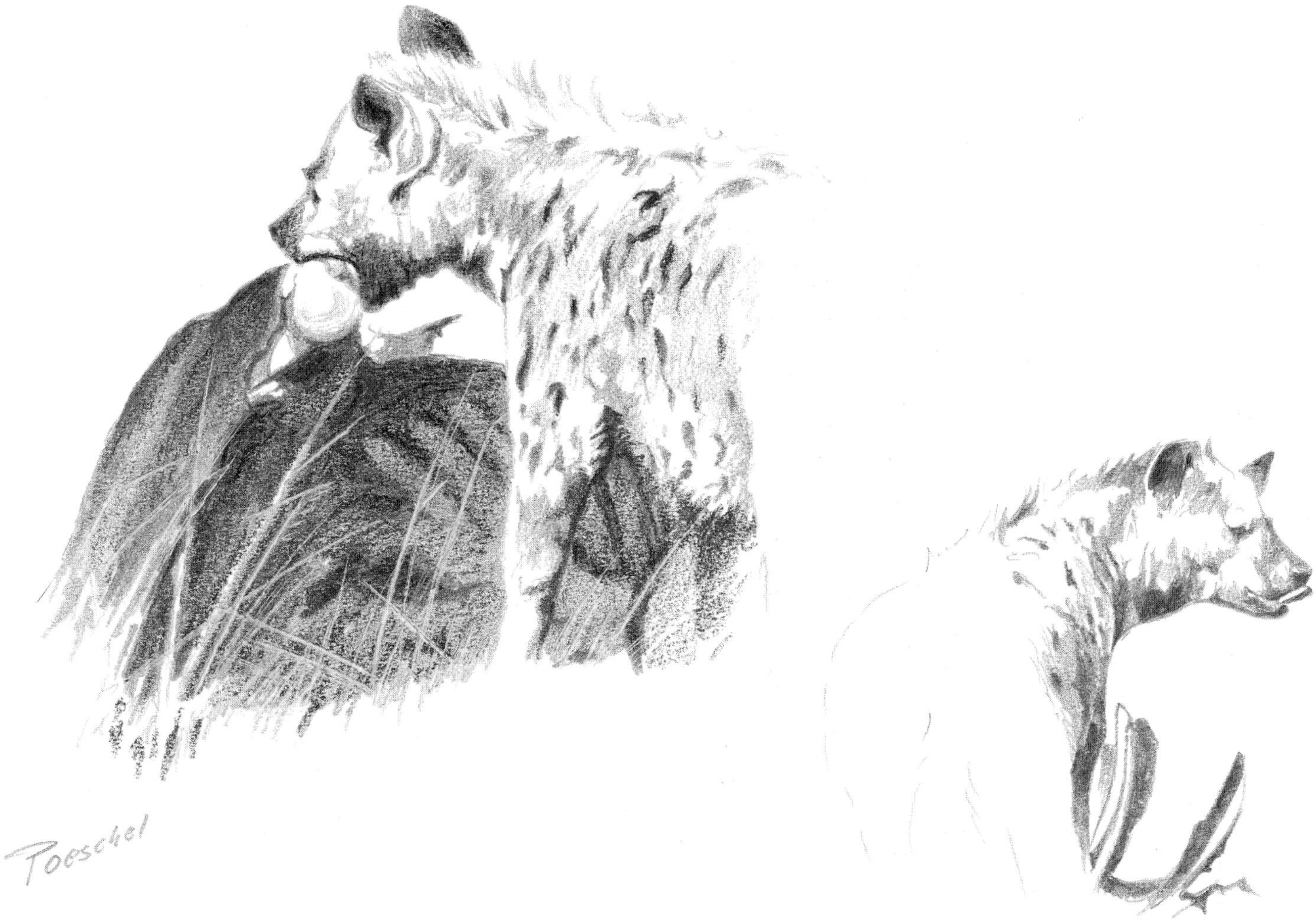
Poeschel

Poeschel

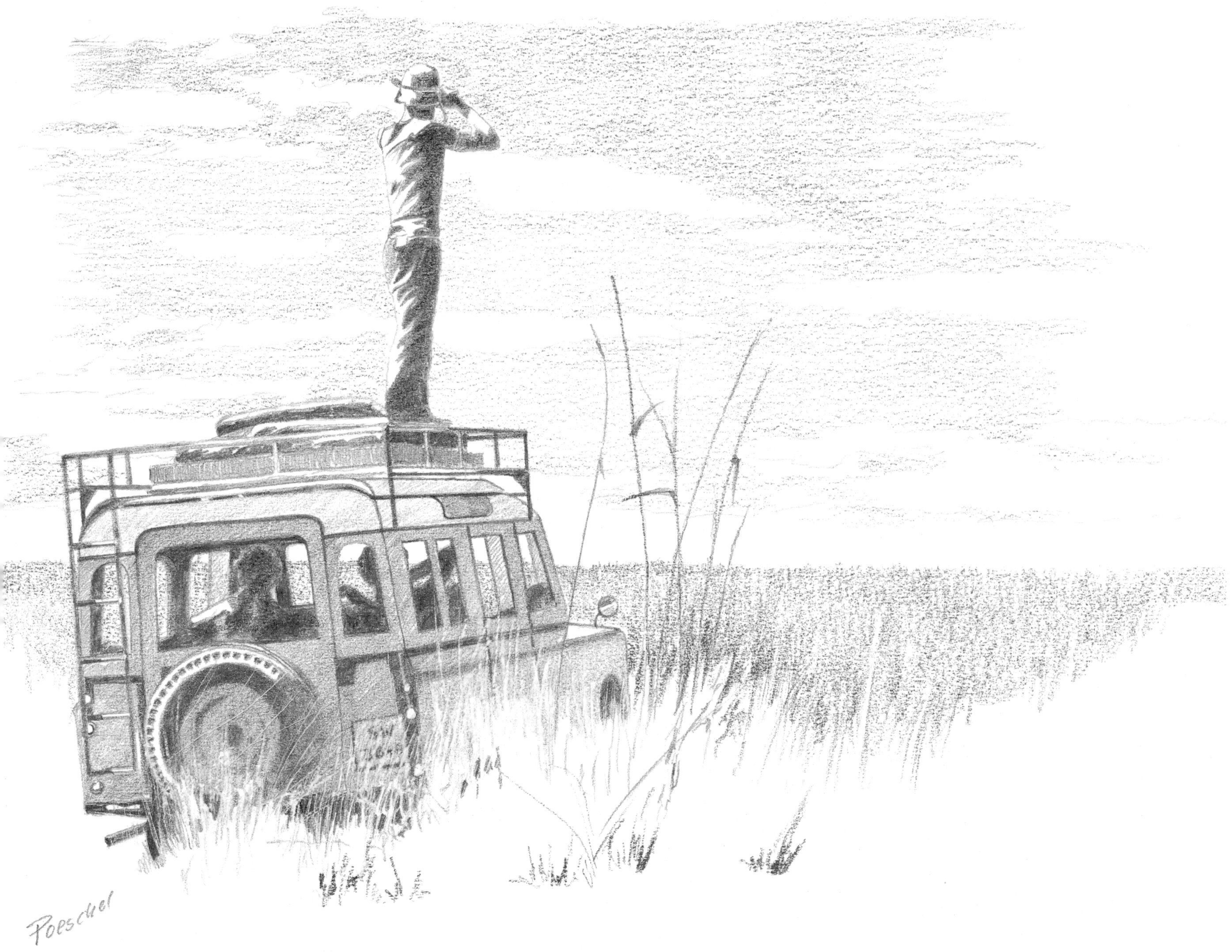
Poeschel

Der Okawango-Fluß entspringt in den Bergen Angolas und fließt zuerst in Richtung des Atlantischen Ozeans, bis Erdverschiebungen ihn zwingen, seinen Weg ins Binnenland einzuschlagen. Schließlich versickert er im 300 m tiefen Sand der Kalahari. Bei nachlassender Kraft der Strömung entstehen fächerförmige Wasserläufe. Diese bilden von Bäumen bestandene Inseln, deren Ufer von Schilf und Papyrus gesäumt sind. Es ist das größte Inland-Delta der Welt mit einer Ausdehnung von 15 000 qkm. Das kristallklare Wasser ist voll von Fischen und auch die Krokodilbestände erholen sich langsam, nachdem die Regierung im Jahre 1975 ein Abschußverbot erlassen hat. Die Krokodile sind zur Regulierung der Fischbestände von großer Wichtigkeit. So stellte man fest, daß sich die Raubfische zu Ungunsten der Friedfische vermehrt hatten.

Die Bezeichnung »Okawango-Sümpfe« ist irreführend, es gibt keine morastigen Sumpfflächen. Die riesigen Tierbestände lockten Jäger, weiße wie schwarze, von überall an. Vor allem die Jagd nach Elfenbein und die rücksichtslose Bejagung dezimierte die Herden. Die ansässigen Eingeborenen vom Stamme der ba Tawana mußten immer weitere Strecken zurücklegen, die Jagd wurde immer schwieriger. Im Jahre 1960 versammelte sich der Stamm, um eine Entscheidung zu fällen. 5000 qkm der traditionellen Jagdgründe wurden zum ersten Wildschutzgebiet des Okawango, dem Moremireservat. Die Regierung Botswanas erklärte bald darauf die aus 1000 qkm Wildnis, Trockensavanne und seichten »Flats« bestehende Häuptlingsinsel »Chief Island« zum Wildreservat. Von Touristen kann dieses Gebiet nur mit dem Flugzeug erreicht werden.

Für die Tierherden der Kalahari ist das Delta von größter Bedeutung. In der Trockenzeit wandern bis zu 12 000 Gnus und Zebras sowie 20 000 Büffel ab, um das üppige Nahrungsangebot des Deltas zu nutzen. Zwar ist das Moremi Wildlife Reserve mit einem ordentlichen Wegenetz ausgestattet, doch sind je nach Wasserstand verschiedene Teile nicht befahrbar. Die einzige Möglichkeit, mit der eigenen Zeltausrüstung inmitten der Tierwelt die Nacht zu verbringen, bieten dem Besucher die zwei Camps »Xuroro« und »Xakanaxa«.

Pläne, den Wasserreichtum des Deltas wirtschaftlich zu nutzen, sind vorhanden. Durch Gutachten und eindringliche Warnungen von Ökologen blieb es jedoch bei Versuchen. Es ist zu wünschen, daß die Vernunft die Oberhand behält, und ein einmaliges Naturwunder der Menschheit erhalten bleibt.

The Okawango River originates in the mountains of Angola and goes first towards the Atlantic Ocean. Later on it is forced by geological shifts to return into the interior of the continent. At the end it oozes away in the sand of the Kalahari, which has a depth of 300 meter. With the diminishing force of the water the river divides into a fan-shaped system of channels, thus forming numerous tree-covered islands. The shores are bordered by reed and papyrus. This is the largest inland delta of the world, extending over 15,000 square kilometers.

The clear water is teeming with fish, and the crocodile population has been regenerating since 1975 when shooting was prohibited. The crocodiles are important for the regulation of the fish stock. It had been recognized that predatory fishes had increased to the disadvantage of those fishes used for nutritional purposes.

The name »Okawango Swamps« is misleading: there are no marshy swamp areas. The enormous game stock attracted hunters, white and black alike, from everywhere. Ruthless hunting decimated the herds. For the local natives of the tribe 'ba Tawana' successful hunting became very difficult. In 1960 the tribe held an assembly to resolve this problem. 5,000 square kilometers of the traditional hunting grounds became the first game reserve of the Okawango, the Moremi Wildlife Reserve. Later on the government of Botswana declared the Chief Island a game reserve, which consists of 1,000 square kiometers of wilderness, dry savannah, and shallow flats. Tourists can reach there only by plane.

The Okawango delta is of great importance for the game herds of the Kalahari. During the dry season up to 12,000 zebras and gnus and 20,000 buffalos migrate into the delta to feed on its luxuriant vegetation. The Moremi Wildlife Reserve has a good road system; several parts are however impassable at high water levels. There are two campsites, Xuroro and Xakanaxa, where the visitor can spend the night amidst all the animal life in his own tent.

There are plans to make use of the rich water reserves of the delta. Expert opinions and warnings by ecologists have, however, prevented efforts like these up to now. It is be hoped that reason will predominate, and that this unique natural miracle will be preserved for mankind.

Moremi Wildlife Reserve
Botswana

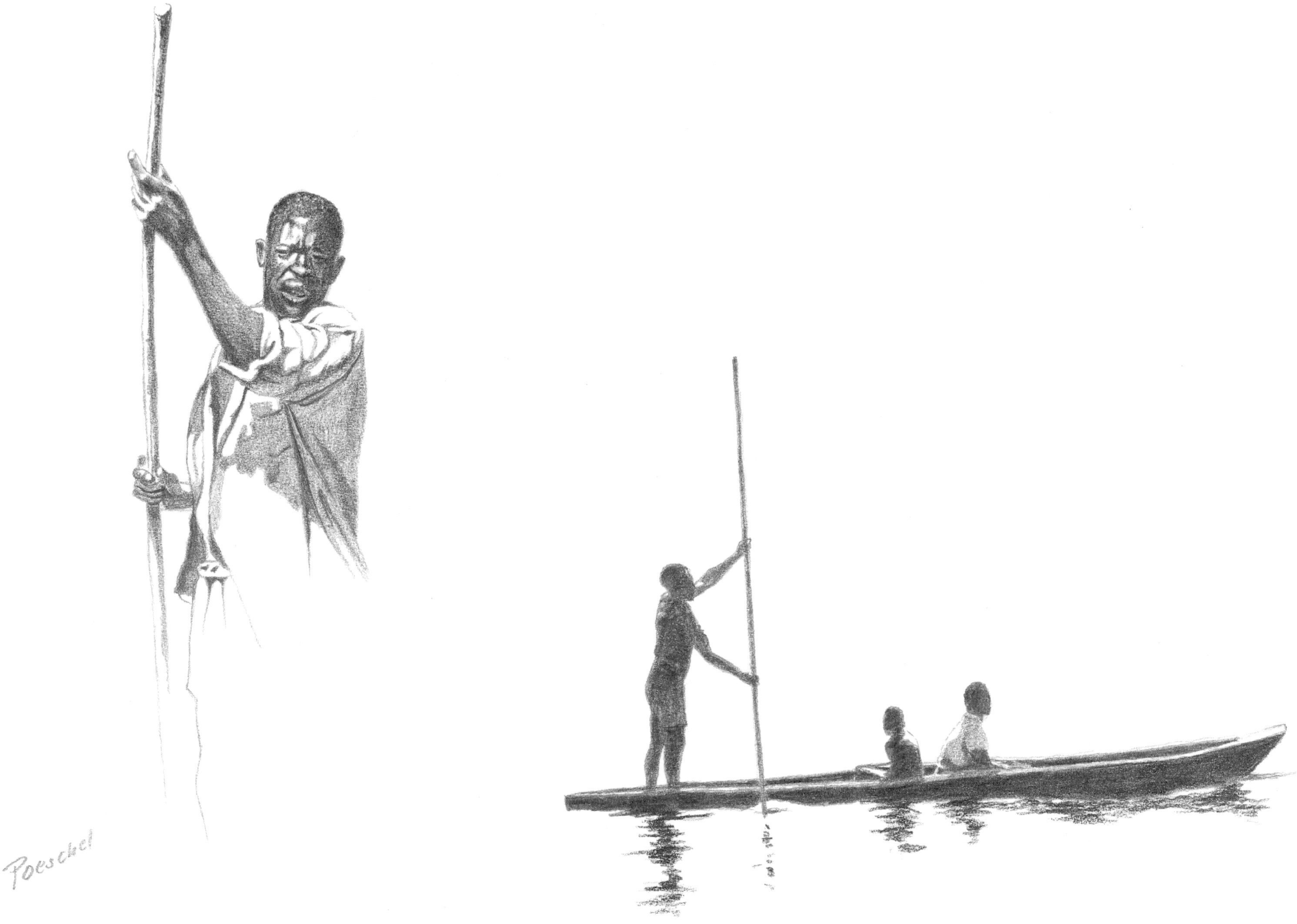
Poeschel

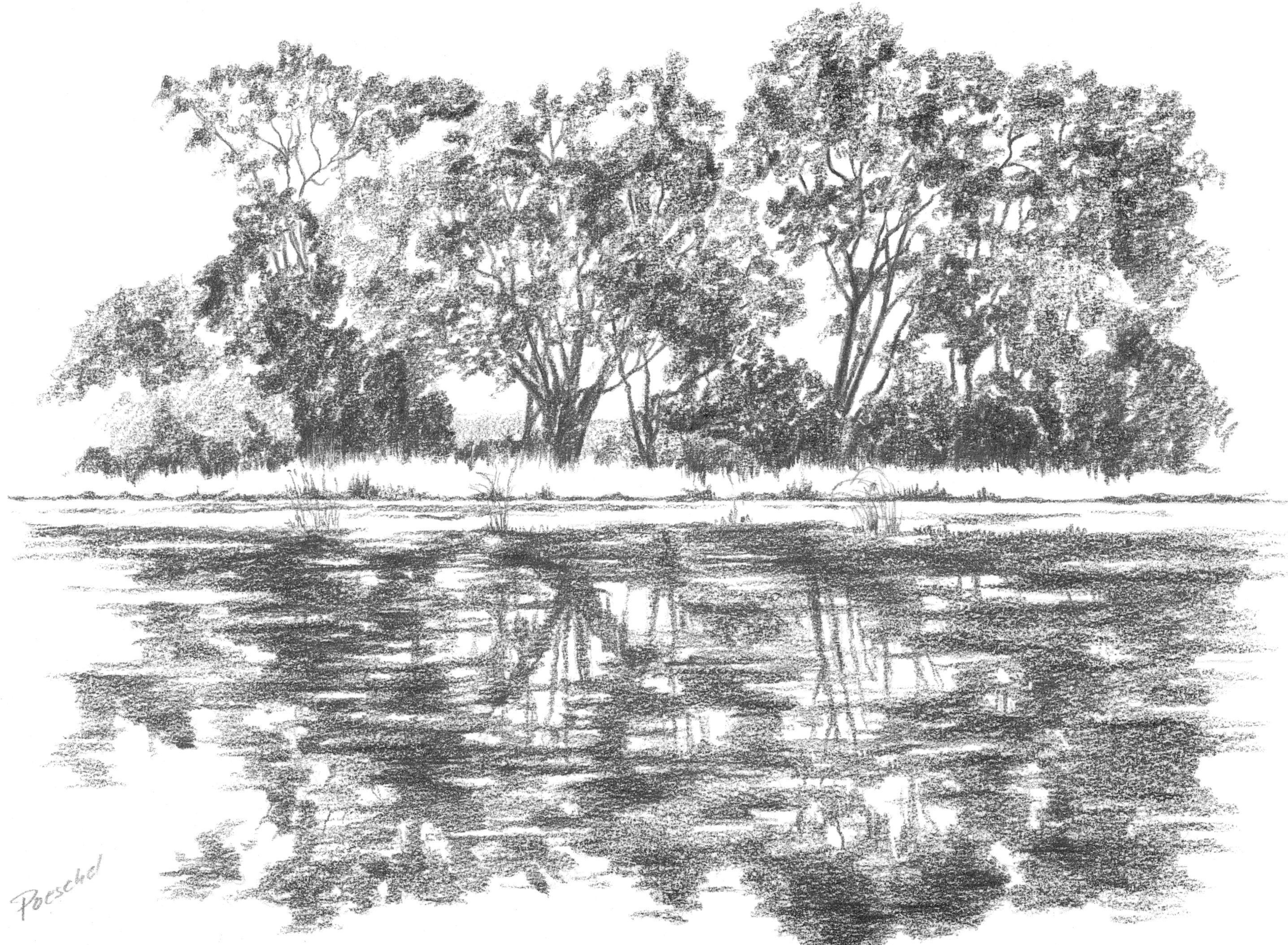
Poeschel

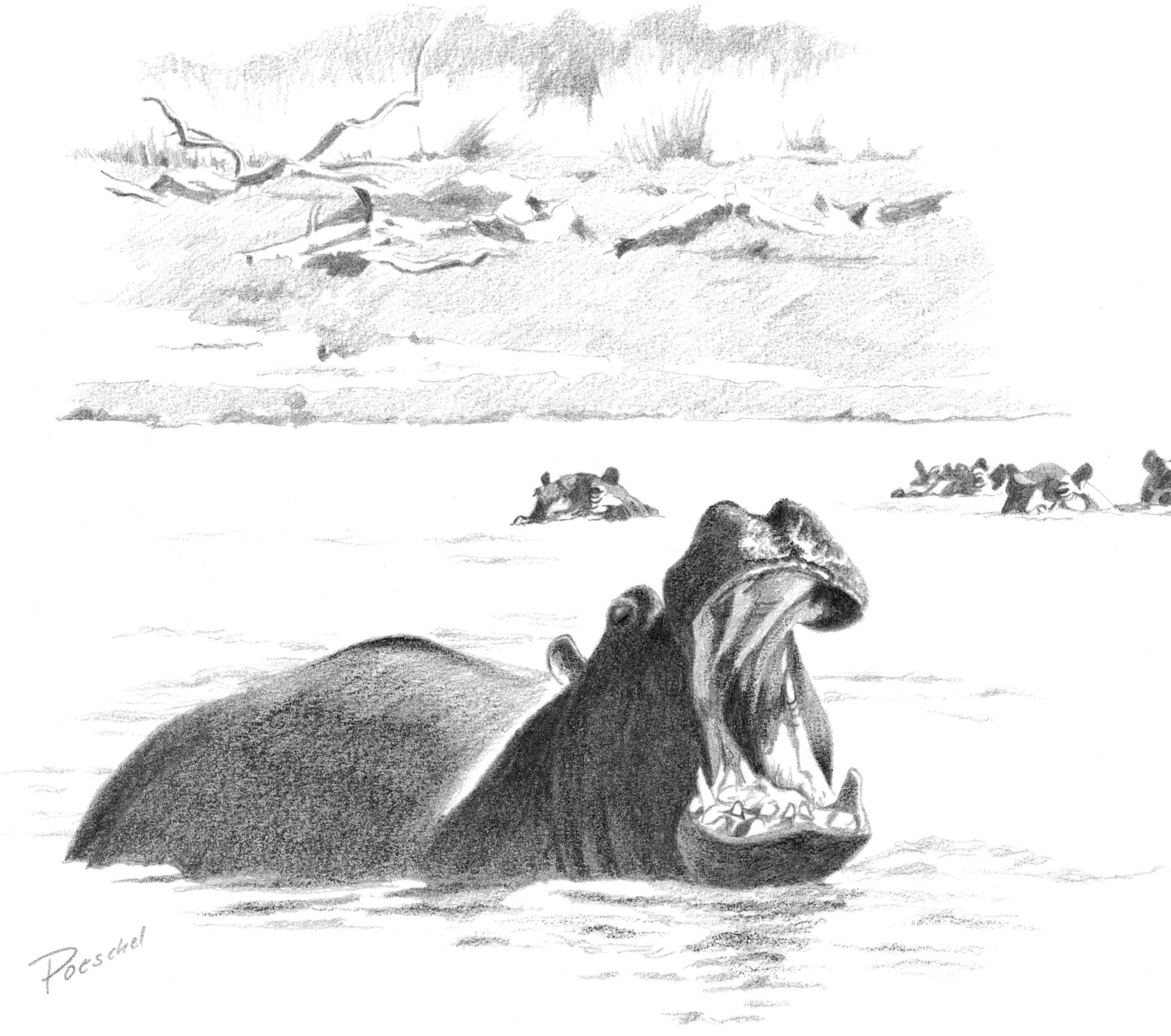
Poeschel

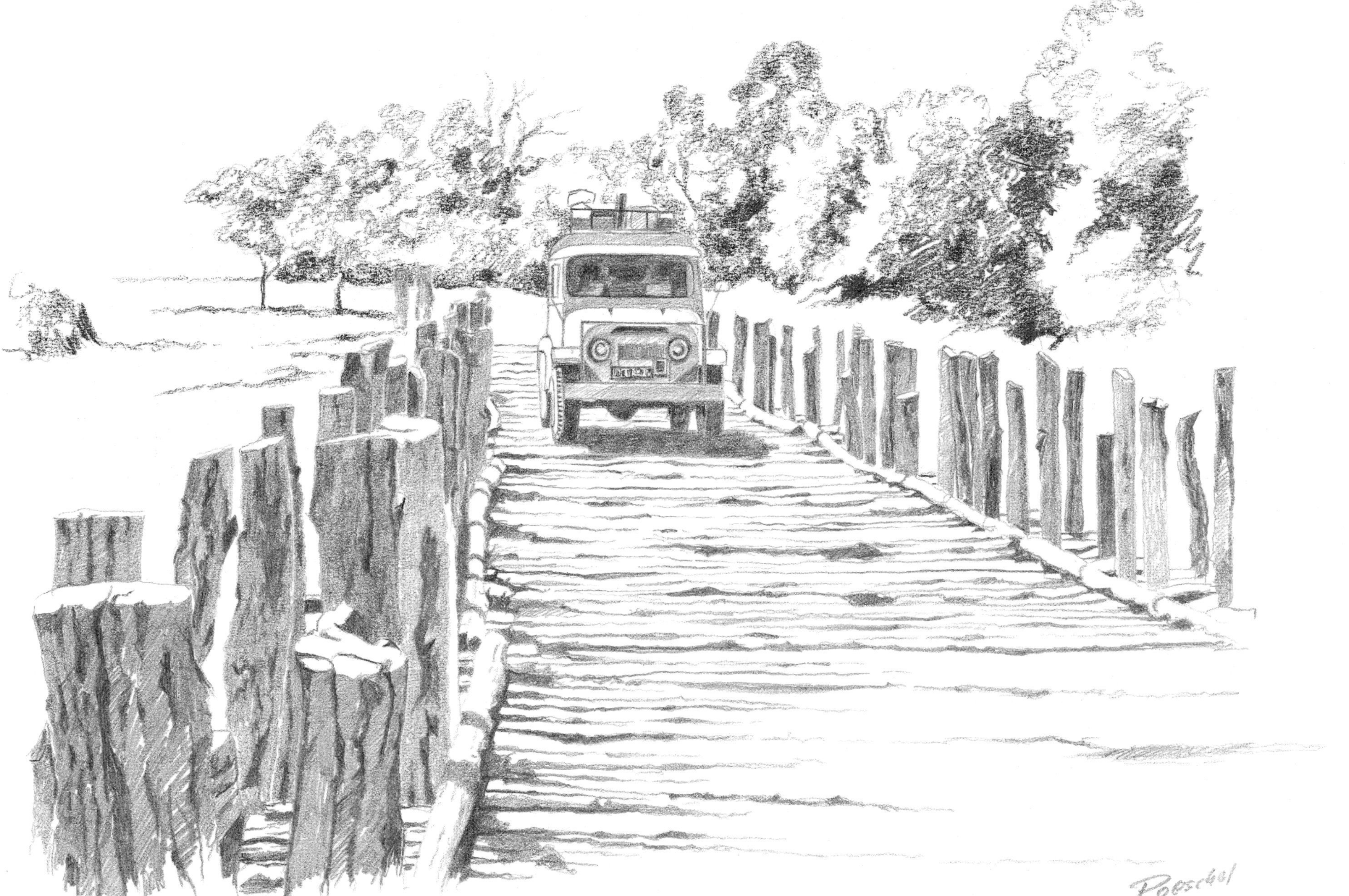

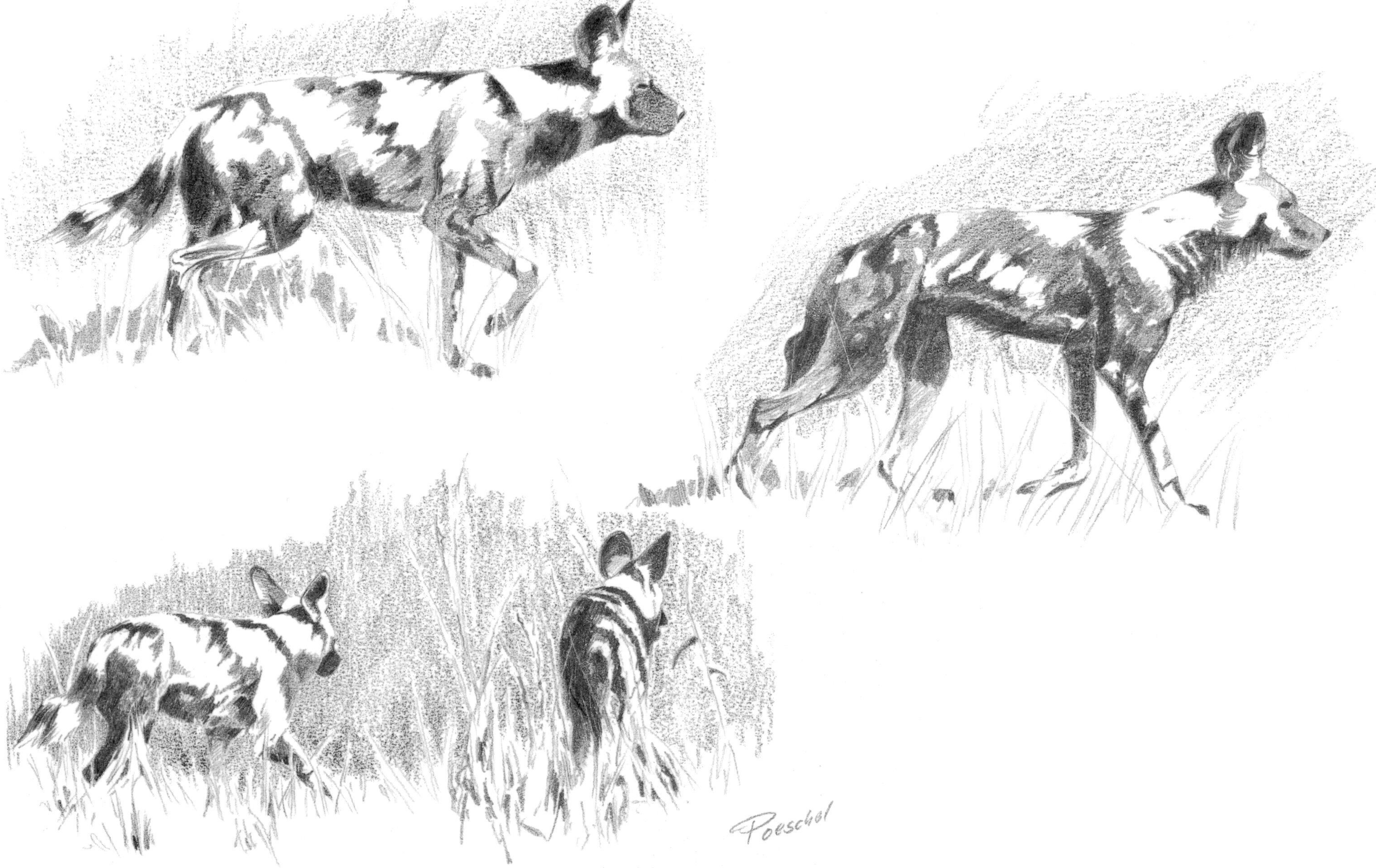
Poeschel

105

Poeschel

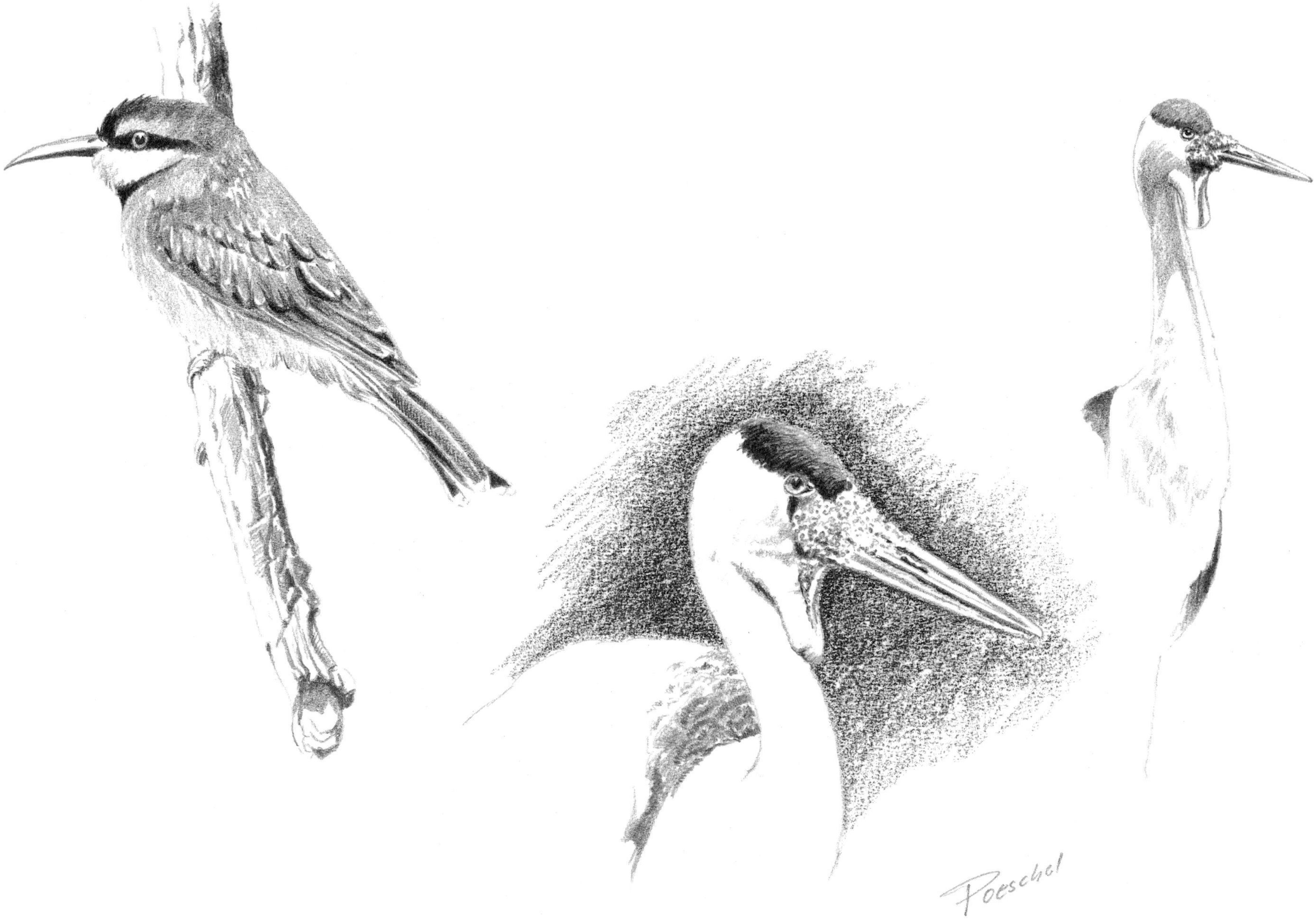
Poeschel

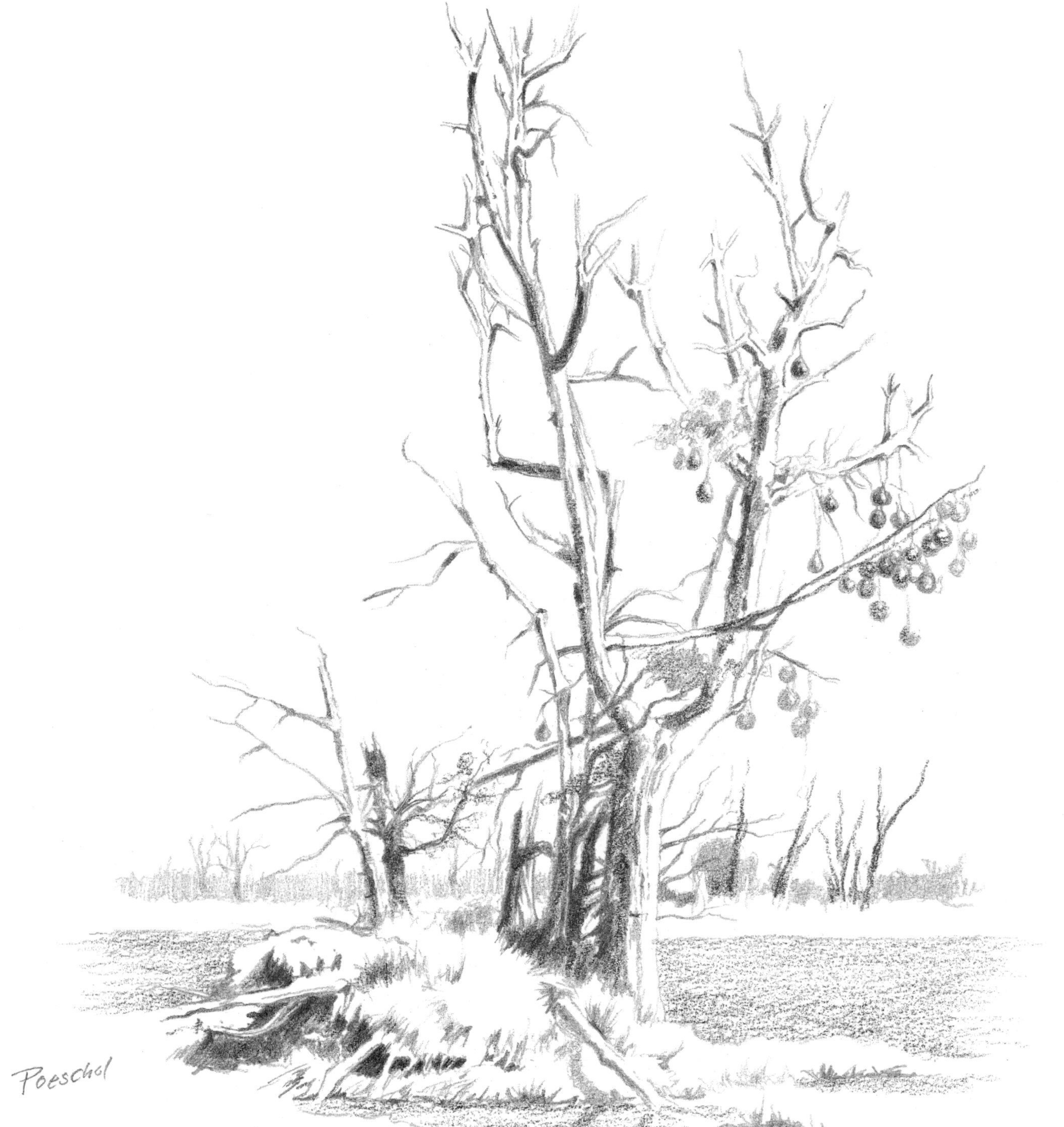
Poeschol

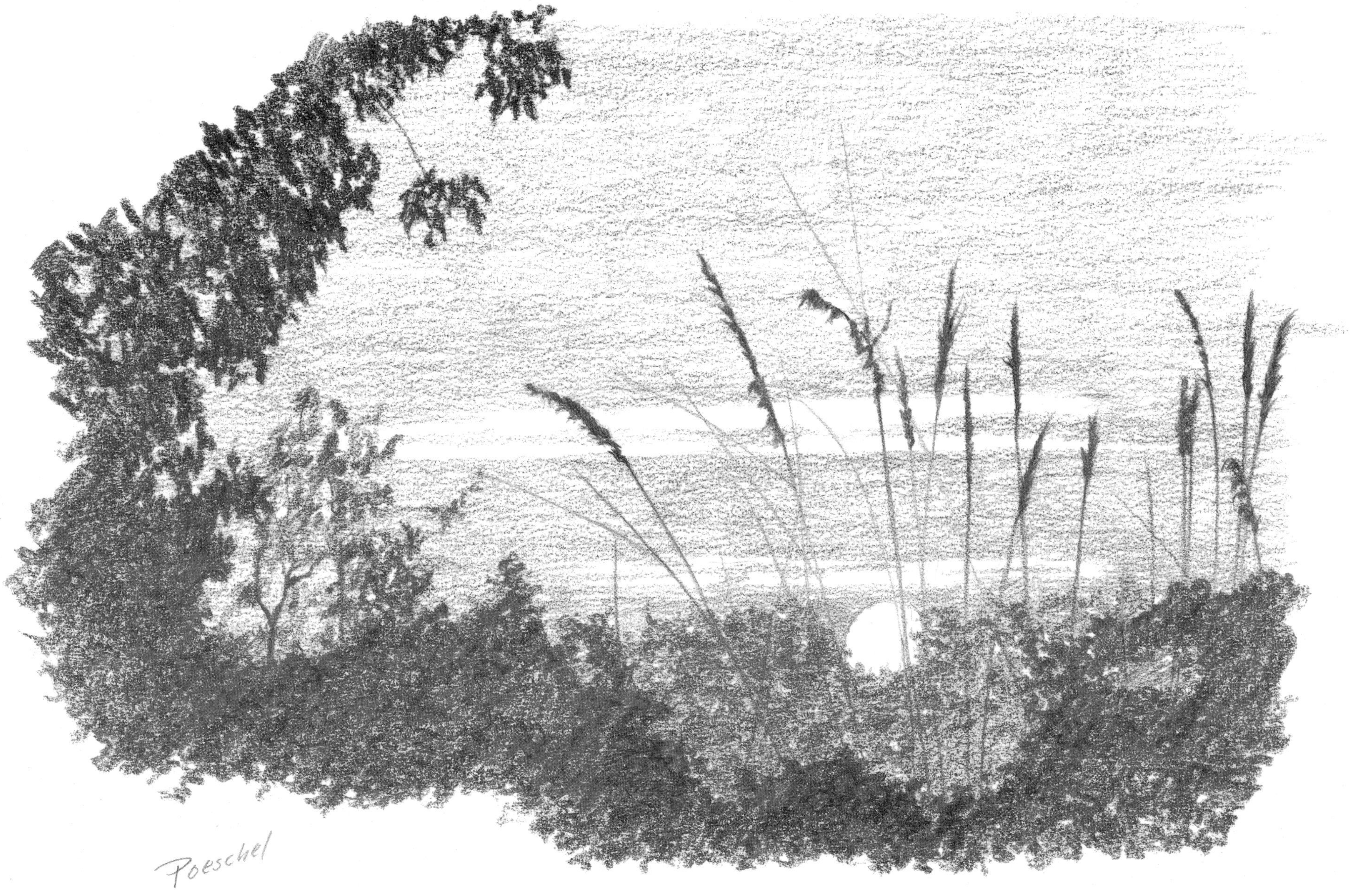
Poeschel

Poeschel

110

Südafrika

1-7 BOPHUTHATSWANA

VENDA

Krüger National Park

TRANSVAAL

4 5 6

O PRETORIA
O Johannesburg

1

2

3

SWAZILAND

ORANJE FREISTAAT

Hluhluwe Game Reserve

Kimberley O

O Bloemfontein

7

NATAL

LESOTHO

TRANSKEI O Durban

KAP - PROVINZ

TRANSKEI

O Kapstadt

O Port Elizabeth

Größtes Wild- und Naturschutzgebiet Südafrikas ist der in Osttransvaal an der Grenze zu Moçambique gelegene Krüger Nationalpark. S.P.J. Krüger, Präsident der Burenrepublik Transvaal, erkannte schon frühzeitig den raschen Niedergang der Wildtierbestände im Buschfeld (Bushveld). Durch seine anhaltenden Bemühungen wurde 1898 zunächst das Gebiet zwischen Krokodil- und Sabie-Fluß zum Wildschutzgebiet proklamiert. Colonel Stevenson-Hamilton setzte weitere Gebietsanschlüsse durch. Mit Unterstützung des Landwirtschaftsminister P.J. Grobler wurde schließlich 1926 der gesamte, fast 20000 qkm große Bereich zum Nationalpark erklärt.

Im westlichen Teil dominieren Granit-, im östlichen Basaltformationen. Das Nyandu-Sandfeld im nördlichen Teil bei Punda Milia weist eine vom übrigen Park abweichende Vegetation auf. Mopane-Wälder beherrschen die Szenerie nördlich des Olifant-Flusses, Eisenholz-, Ebenholz- und Mahagonibäume die Ufer des Luvhuvhu-Flusses. Dichtes Buschland und Akazienbestände breiten sich vom Rastlager Skukuza bis Lower Sabie. Die Höhenunterschiede liegen zwischen 200 m im Osten und um 900 m bei Pretoriuskop.

Eine Zählung der Tierbestände im Jahre 1983 ergab Kopfzahlen von 8000 Elefanten, 29000 Büffel, 25000 Zebras und 153000 Impalas, um nur die größten Populationen zu nennen. Erwähnenswert ist auch die reichhaltige Vogelwelt, mehr als 400 Arten wurden registriert.

Der Krüger ist der meistbesuchte Nationalpark Afrikas. 1977 wurden 100000 Fahrzeuge und 377000 Besucher gezählt. Davon kommt die Mehrzahl aus den größeren Städten Südafrikas mittels organisierter Busreisen. Bei solch hohem Touristenaufkommen ist der Aufenthalt im Schutzgebiet strengen Reglements unterworfen. Es ist weder gestattet, mit dem Fahrzeug die Wege, noch außerhalb der erlaubten Plätze das Fahrzeug selbst zu verlassen. Die Wildhüter achten unnachsichtig auf Einhaltung der Bestimmungen. Eine Besonderheit sind sogenannte »Wilderness Trails«. Gruppen bis maximal 8 Personen können, von einem erfahrenen Wildhüter geführt, Fauna und Flora zu Fuß erkunden.

The largest game and nature reserve of South Africa is the Kruger National Park in eastern Transvaal, adjacent to the border of Mozambique. S.P.J. Kruger, president of the Boer republic of Transvaal, recognized in time the fast decline of the game stocks in the Bushveld. Due to his continuous efforts the area between the Crocodile river and the Sabie river was declared a game reserve in 1898.

Colonel Stevenson-Hamilton succeeded in enlarging this area. Finally in 1926, with the support of Minister of Agriculture P.J. Grobler the whole area included today, comprising nearly 20,000 square kilometers, was declared a national park.

In the western part granites are the prevailing geological formations, while in the eastern part basaltic rocks dominate. The Nyandu sand flat in the northern part near Punda Milia has a vegetation that differs from the remainder of the park. Mopane forest dominates the landscape north of the Olifant river, the Luvhuvhu river is bordered by ebony and mahagony trees. Dense bush and acacia stands spread from Camp Skukuza to Lower Sabie. The elevation varies between 200 meters in the east to 900 meters near Pretoriuskop.

A game census in 1983 amounted to 8,000 elephants, 29,000 buffalos, 25,000 zebras and 153,000 impalas, if we consider the species with the highest population numbers.

The park contains a rich bird life, more than 400 species have been counted.

The Kruger Park has the highest number of visitors of all African National Parks. In 1977 100,000 cars and 377,000 visitors were registered. Most of these come by organized bus tours from the bigger towns of South Africa. With such a high number of tourists travelling is subject to strong regulations. Visitors are not allowed to leave the roads with the car nor to get out of the car outside the camps and places especially indicated. The game wardens ensure strict observance of the regulations. A highlight are so-called wilderness trails. Guided by an experienced game warden groups of up to eight persons can explore fauna and flora on foot.

Kruger National Park
Südafrika

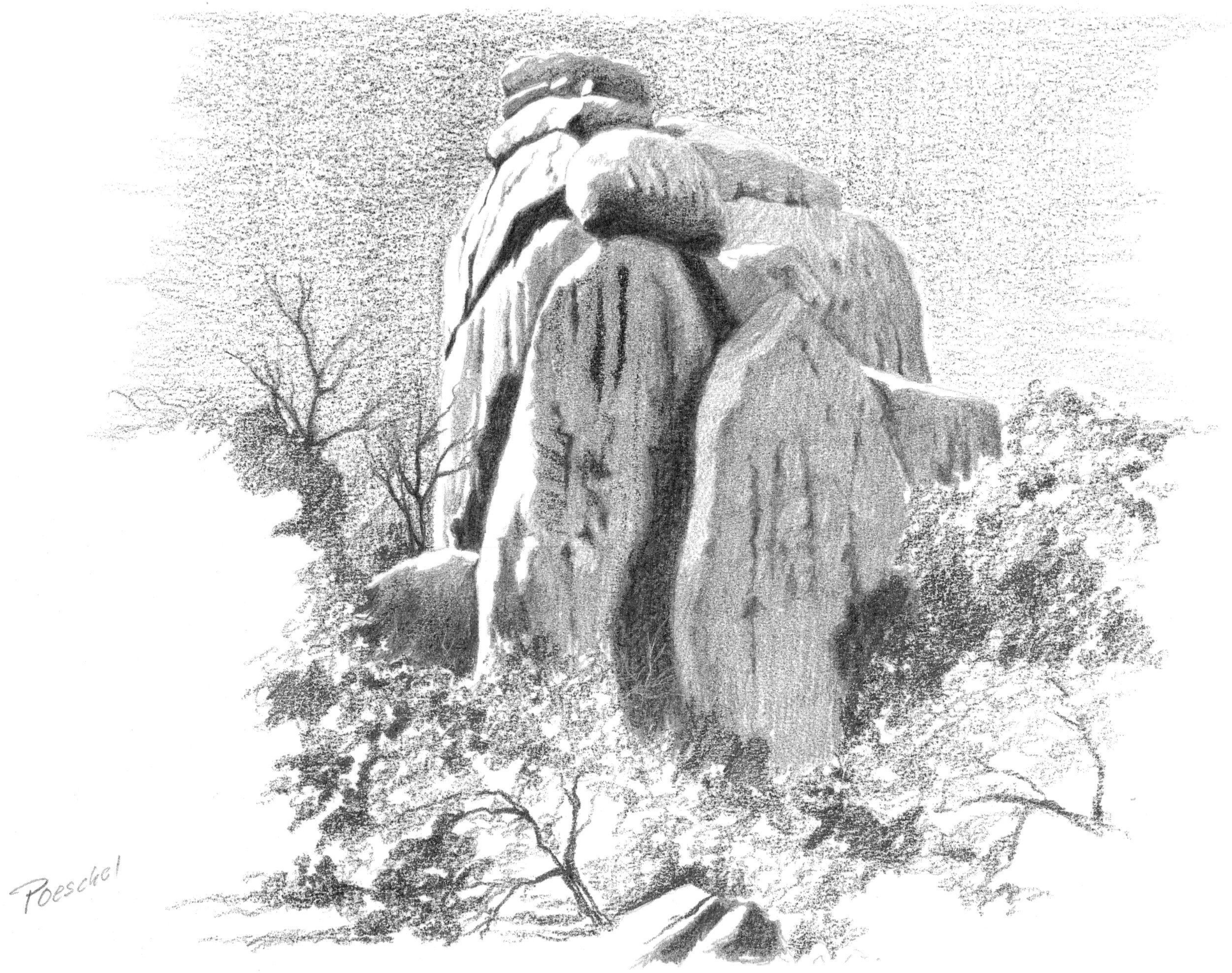
Poeschel

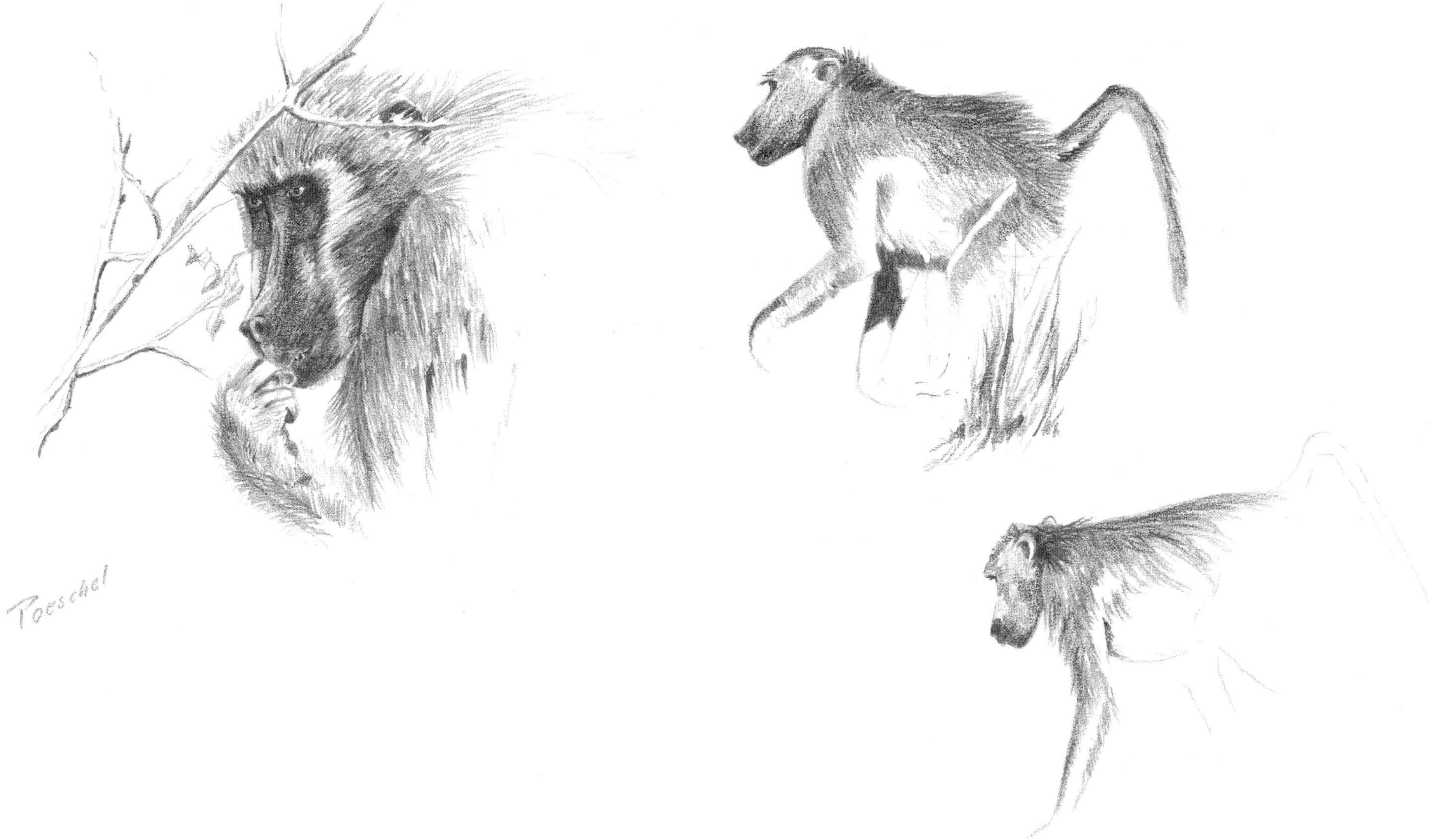
Poeschel

♀

Poeschel

Poeschel

♂

♀

Poeschel

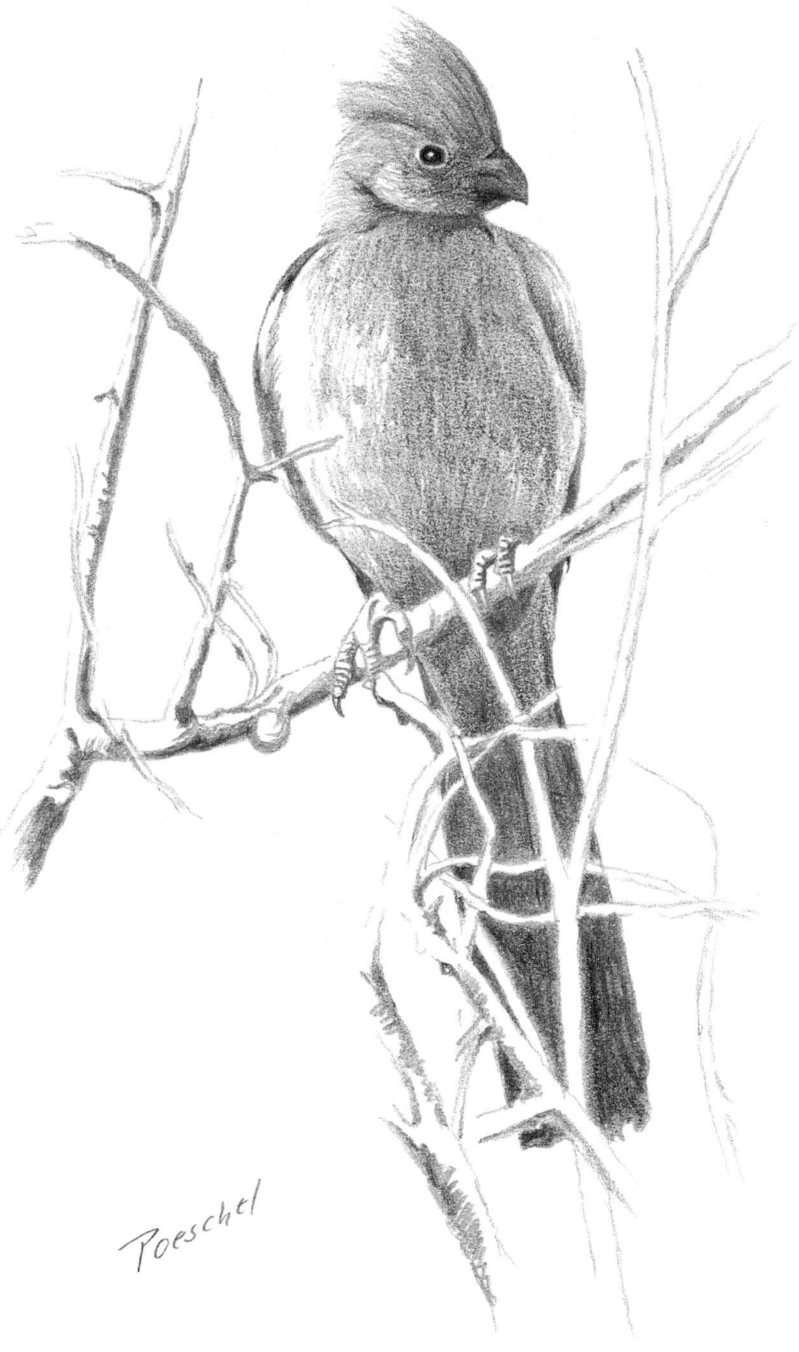
Poeschel

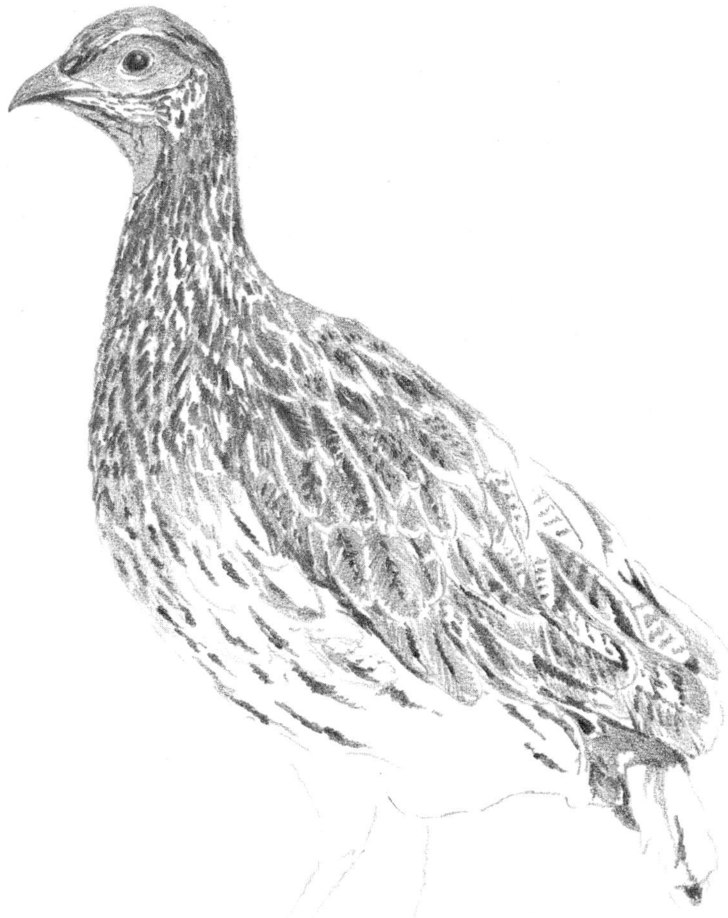

Poeschel

122

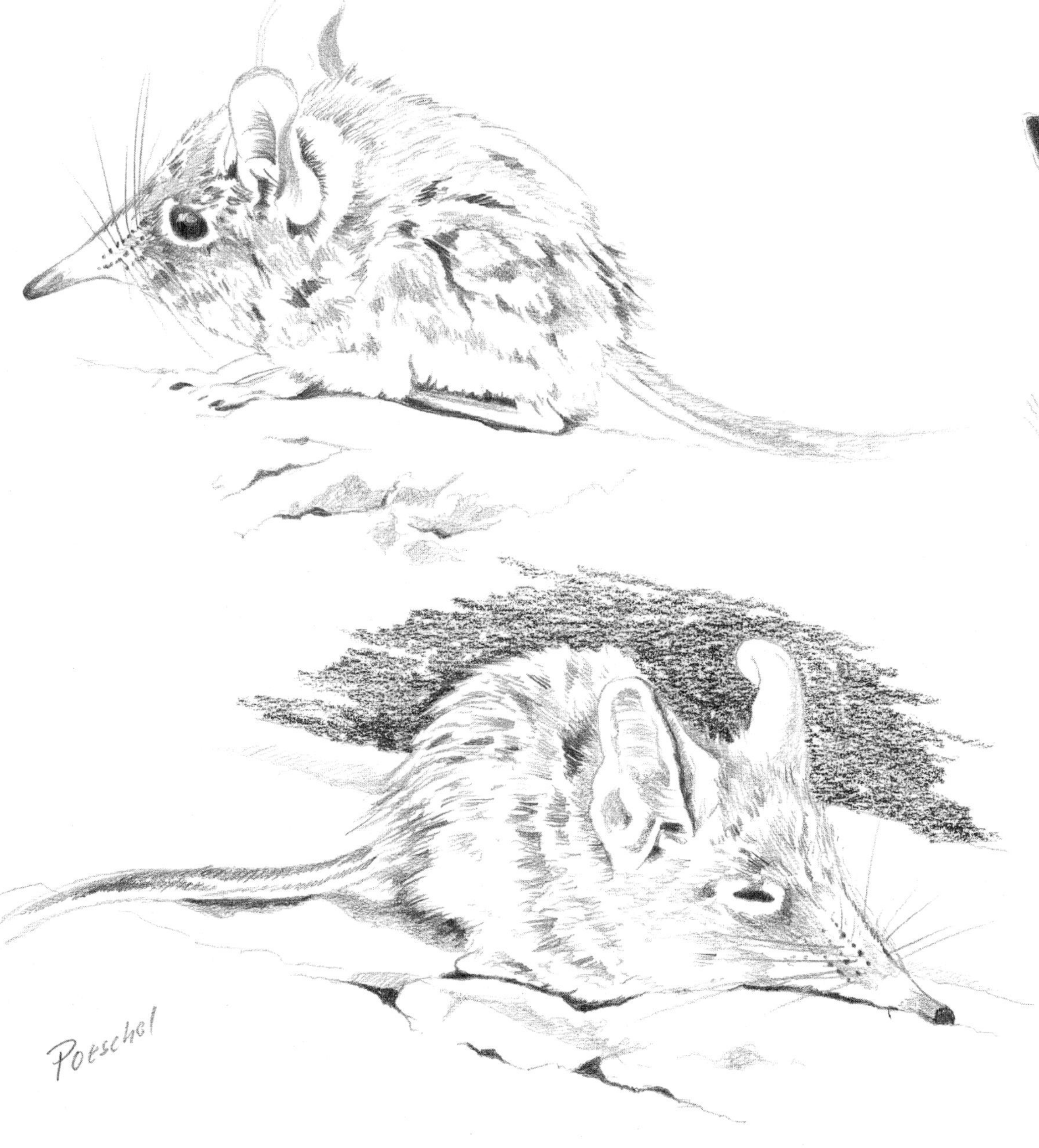
Poeschel

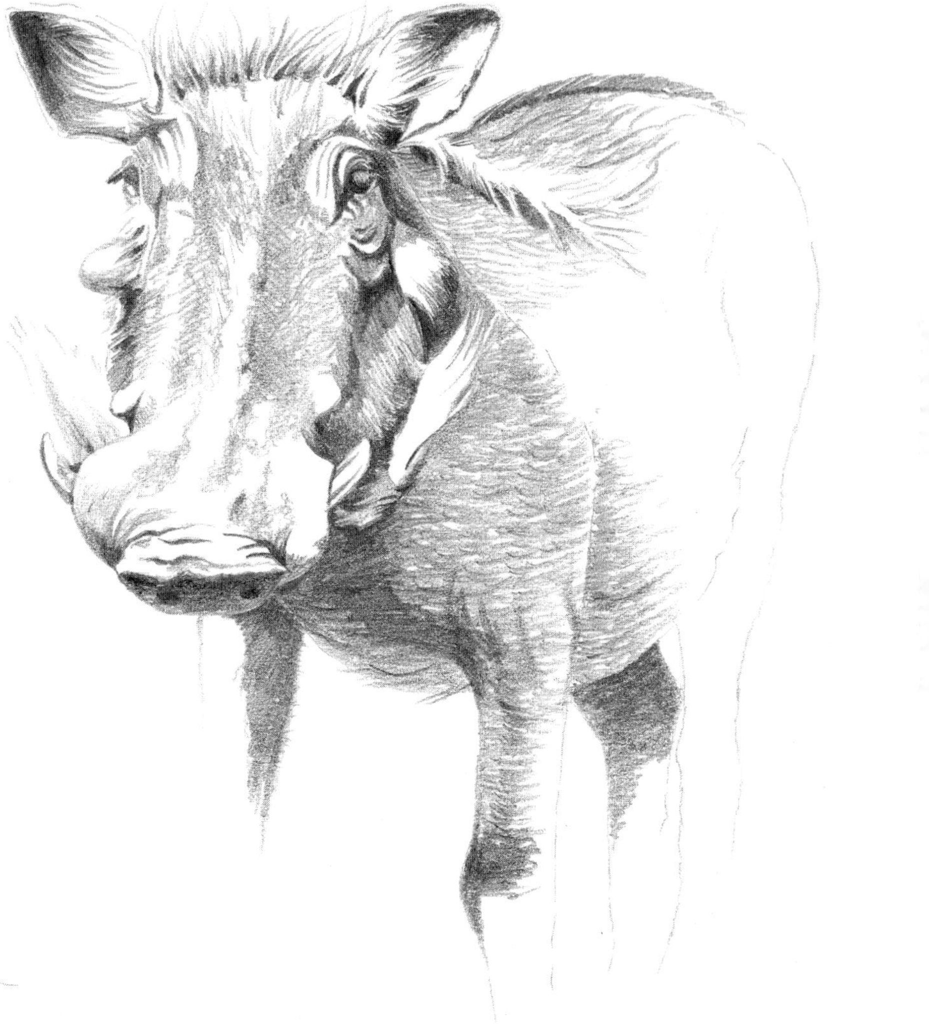

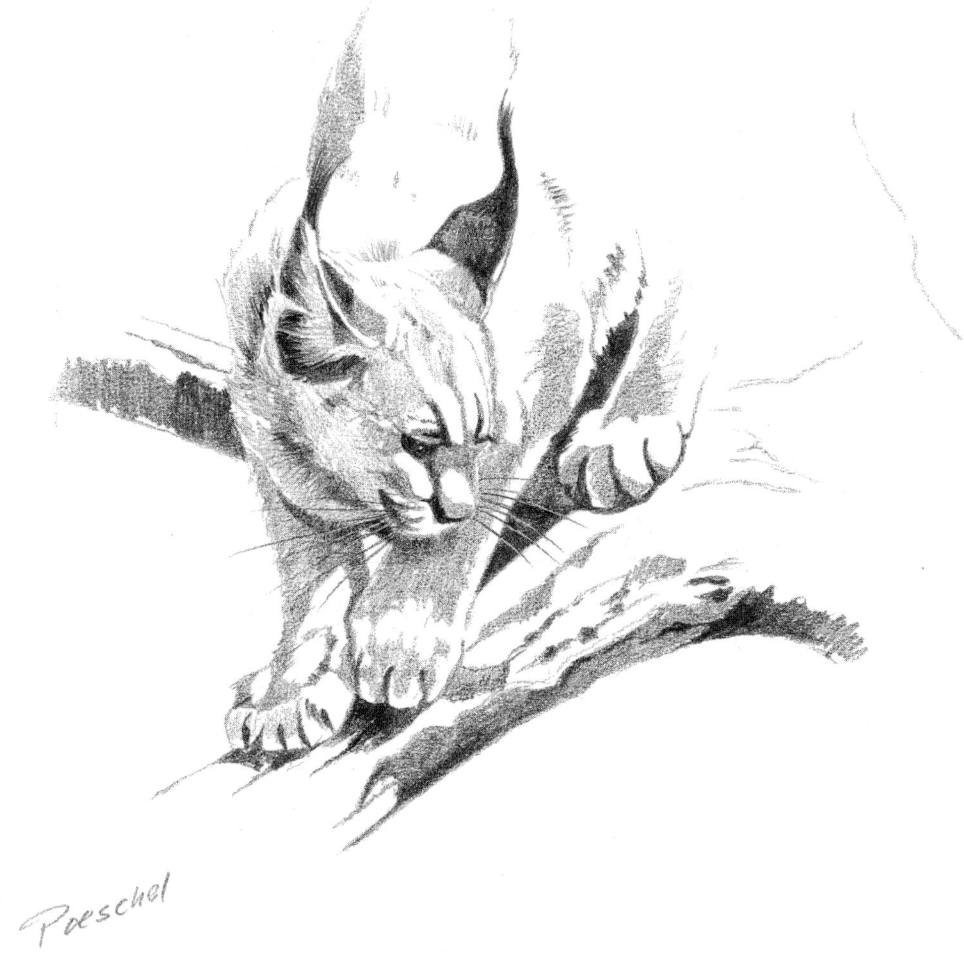
Poeschel

♀

♂

Poeschel

125

Das Wildschutzgebiet Hluhluwe wurde 1897 zusammen mit Umfolozi und St. Lucia gegründet. Der Name wird von einer Kletterpflanze abgeleitet, dem in Flußgebieten gedeihenden »Dornigen Affenseil« (Dalbergia armatal), in der Zulusprache »umhluhluwe« genannt. Das Wildreservat ist 23000 ha groß, hügelig, und liegt zwischen 80 und 600 m über dem Meeresspiegel. Parkähnliche Savannen wechseln mit Busch- und offenem Grasland.

Den strengen Schutzbestimmungen der Wildschutzbehörden in Südafrika ist es zu verdanken, daß das »Weiße« oder Breitlippen-Nashorn nicht nur überlebte, sondern heute wieder in beachtlichen Beständen vorhanden ist. Im Umfolozi sind es allein 900 Exemplare. Auf Grund der Vermehrung konnten Tiere in andere Reservate übersiedelt und auch an Zoos abgegeben werden. Die nördlich des Äquators lebende Unterart dürfte kurz vor der Ausrottung stehen. Das Breitlippen-Nashorn ist nach dem Elefanten das zweitgrößte Landsäugetier und ist von friedlicher Natur. Beide afrikanischen Nashornarten kommen nebeneinander vor, sie sind keine Nahrungskonkurrenten. Das »Weiße« Nashorn ist Grasesser, das »Schwarze« zupft mit seiner verlängerten Oberlippe Blätter von Büschen. Die Nashornbestände sind drastisch zurückgegangen. Schuld daran ist die große Nachfrage asiatischer Länder. Dort findet das pulverisierte Horn als »Arznei« und Aphrodisiakum reißenden Absatz. Der zweitgrößte Abnehmer ist Nord-Jemen, wo es zur Manneszierde gereicht, einen Dolch mit Griff aus Nasenhorn zu tragen.

In den Tierreservaten Natals gibt es für den Besucher »Verstecke«, harmonisch auf Pfählen in die Landschaft eingefügte Beobachtungsstände, von denen man Tiere an den Wasserstellen beobachten kann. Besonders erwähnenswert ist die Nyala-Antilope. Das männliche Tier ist durch seine Rücken- und Halsmähne, die weißen Seitenstreifen und ockerfarbenen Beine besonders auffällig gezeichnet. Auch leben von dem fast ausgerotteten Burchell-Zebra kleine Gruppen im Park. Die Unterbringung ist wie in allen südafrikanischen Wildreservaten hervorragend und preiswert.

The game reserve Hluhluwe was founded in 1897, together with Umfolozi and St. Lucia. The name is derived from a creeper, the »Thorny Monkey's Rope« (Dalbergia armatal), which thrives at the river banks and is called »umhluhluwe« in the Zulu language. This game reserve has an extension of 230 square kilometers and is situated in a hilly landscape between 80 and 600 meters above sea level. Park-like savannah alternates with bush and open grassland.

Owing to the strict regulations of the South African game conservation authorities the White Rhino has not only survived but increased to considerable stock again. The Umfolozi alone is today populated by 900 White Rhinos. Due to a good propagation rate a number of animals could be transferred to other parks or given to zoos. On the other hand the subspecies living north of the equator is probably seriously threatened by extinction. The White Rhino is after the elephants the largest living land mammal, and as a rule it is a peaceful creature. Both African species of rhino can live in the same habitat, they are no food competitors. The White Rhino feeds on grass, the Black Rhino has a specialized prehensile upper lip to pluck leaves and twigs off bushes and trees. The numbers of rhino have drastically declined in recent times, due to the strong demand for the pulverized horns from Asiatic countries. This is highly esteemed, and highly paid for, as an aphrodisiac. The second largest customer is North Yemen, where it is a symbol of virility to possess a dagger with a grip of rhino's horn.

In the game reserves of Natal there are hides for visitors, observations posts on stakes, from where the game can be watched at watering places. Especially remarkable is the Nyala Antilope. The male is a conspicuous animal with a mane on neck and back, white stripes on flanks and ochre legs. The nearly extinguished Burchell's Zebra lives in small parties in the park. The accomodation is, as in all South African game reserves, excellent and comparably cheap.

Hhuhluwe Game Reserve
Südafrika

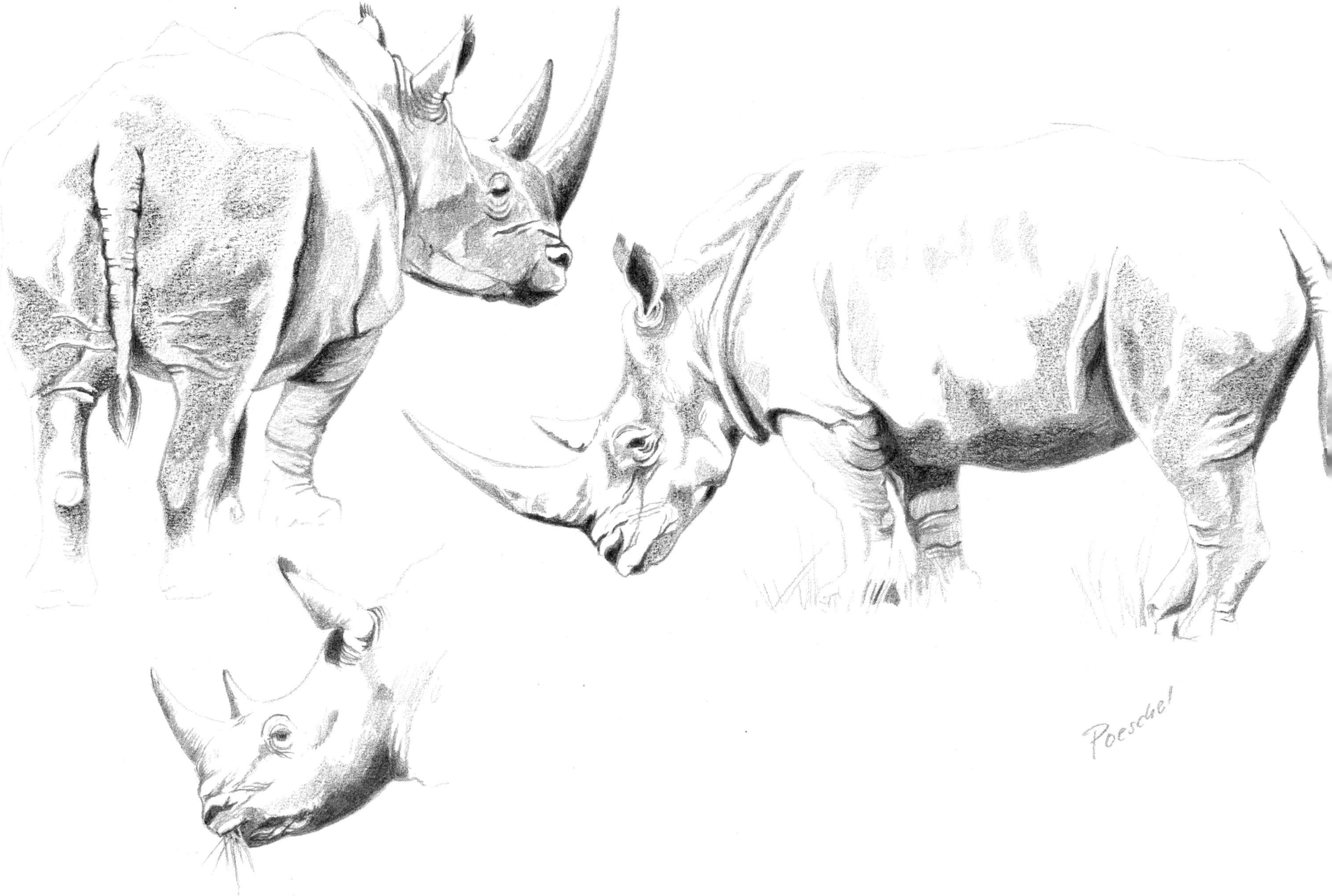
Poeschel

Poeschel

Poeschel

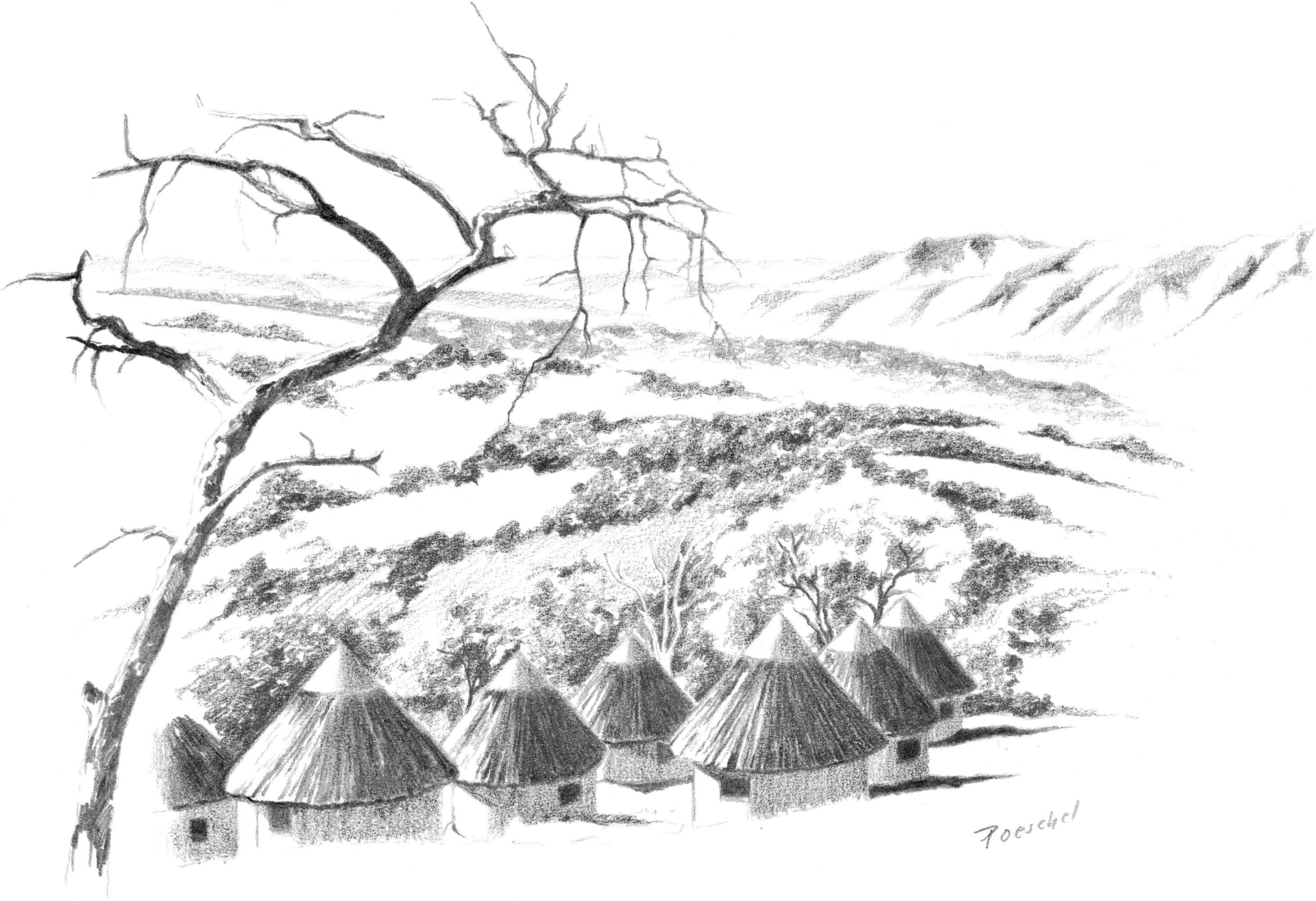
Poeschel

♂

♂

♀

Poeschel

♂

Poeschel

Poeschel

Poeschel

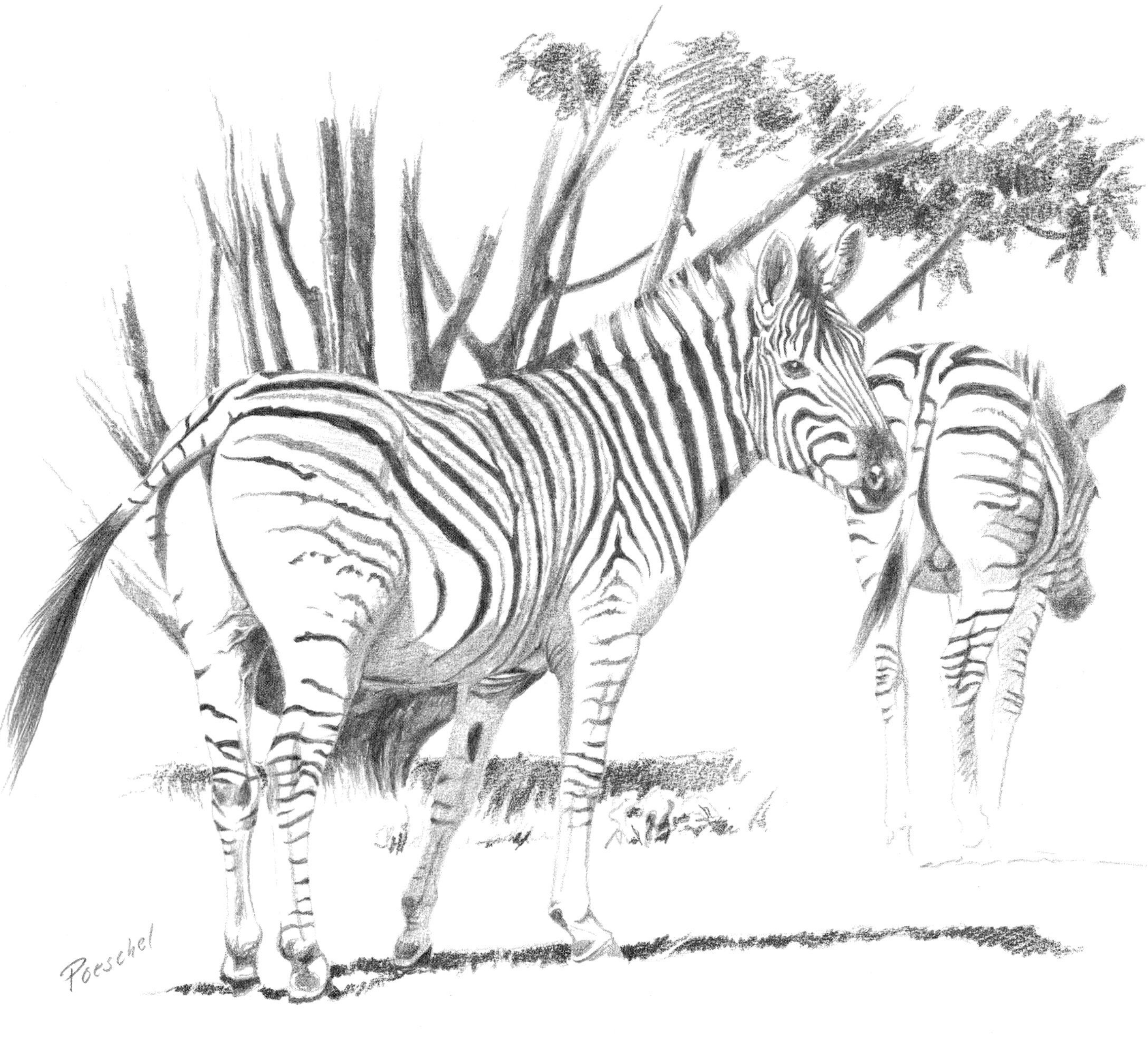
Poeschel

Poeschel

137

Poeschel

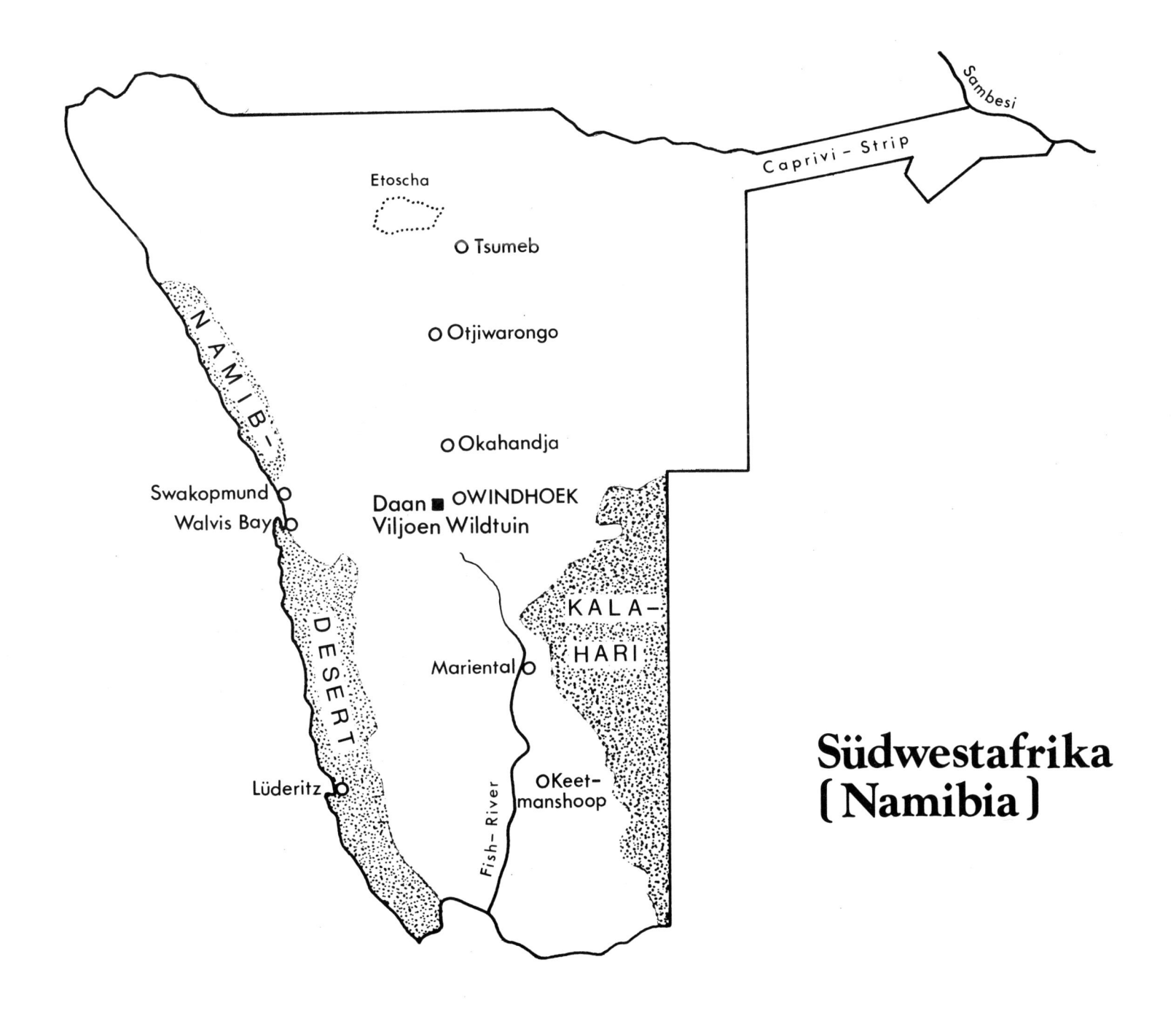

Südwestafrika (Namibia)

24 km westlich von Windhuk (der Hauptstadt Südwestafrikas) liegt inmitten der welligen Hügellandschaft des Khomas-Hochlandes das Daan-Viljoen-Wildschutzgebiet.

Es wurde im Jahre 1962 durch die Initiative des damaligen Administrators D.T. du P. Viljoen zum Wildreservat proklamiert. Es ist 4000 ha groß und für die Windhuker eine Erholungsstätte am Wochenende. Es liegt 2000 m über dem Meeresspiegel und bildet mit seinem zerklüfteten Bergland, mächtigen Felswänden und Schluchten eine rauhe und herbe Landschaft. Es wird im Westen von der Namibwüste und im Osten von der Windhuker Talsohle begrenzt. Zentrum des Parks ist der Augeigasdamm mit Rastlager, ein Standort, der gute Möglichkeiten zur Vogelbeobachtung bietet und von dem schon 200 verschiedene Arten registriert wurden. Er beherbergt außerdem die für Südwestafrika typischen Wildarten wie Elen-, Oryx-und Kuhantilopen (Kaamas), Gnus, Kudus, Springböcke, Impalas und die seltenen Bergzebras, sowie Paviane und Strauße, um nur die wichtigsten zu nennen.

Das Besondere im Daan-Viljoen-Park ist, daß man Fußwanderungen ohne Begleitung eines Wildwarts machen kann, denn es gibt keine Raubtiere. Von dieser Möglichkeit habe ich während meines mehrtägigen Besuches im Jahre 1981 reichlich Gebrauch gemacht. Allerdings traf ich eine Situation an, welche den Publikationen und Informationen der einschlägigen Literatur völlig entgegengesetzt war. Sämtliche Staudämme waren leer, das Land völlig ausgetrocknet, es hatte schon seit mehreren Jahren nicht mehr geregnet. Trotzdem erschlossen mir die Wanderungen unbeschreibliche Schönheiten dieser herben Landschaft. Sehr beeindruckend ist die Vegetation und besonders der Charakterbaum Südwests, der Kameldorn, aber auch die landschaftsprägenden Aloen und Euphorbien sind zu erwähnen. Eine Pirsch zu Fuß ist besonders reizvoll, da die Tiere eine normale Fluchtdistanz haben. Eine Herde Kaamas, etwa 50 Tiere, konnte ich einen längeren Zeitraum ungestört beobachten, ebenso die artenreiche Vogelwelt am Rastlager.

Wenn dieser Park auch nicht mit den großen Nationalparks vergleichbar ist, so vermittelt er doch vielseitige Eindrücke einer interessanten Fauna und Flora mit dem Hintergrund einer imposanten Landschaftskulisse.

Twenty four kilometers west of Windhoek, the capital of South West Africa, in the middle of the hilly landscape of the Khomas Highland is the Daan Viljoen Wildlife Reserve. It was declared a Wildlife Reserve in 1962 through an initiative of the administrator D.T. du P. Viljoen. It extends over 40 square kilometers and is a weekend recreation area for the inhabitants of Windhoek. It is situated at an elevation of 2000 m, in a rough mountainous area with steep rocky mountain-sides and ravines. To the west it is bordered by the Namib desert, to the east by the valley of Windhoek. The centre of the park is the Augeigas Dam where there is a resting camp, and from where good possibilities for bird-watching exist. Already 200 bird species have been recorded there.

Numerous game species live in the park, such as Eland, Oryx and Kaama Antelope, Gnu, Kudu, Springbuck, Impala and the rare Mountain Zebra, baboons and Ostrich, to name only the most important.

In the Daan Viljoen Park one is allowed to walk without being accompanied by a game warden, since there are no big carnivores. During my visit in 1981 I took full advantage of this freedom. The situation in the park was however quite different from what I had learned from published literature and other sources. The region was completely dry and the dams were empty since it had not rained for years. Despite this the hiking tours were fascinating because of the indescribable beauty of the landscape and the impressive vegetation including the characteristic tree of South West Africa, the Camel Thorn, and the conspicuous Aloe and Euphorbia bushes. I watched a herd of Kaama Antelope undisturbed for quite a long time, and saw many bird species in the resting camp.

Even if this park cannot be compared to the large national parks, it gives a wide range of impressions of an interesting fauna and flora, on the background of a beautiful landscape.

Daan Viljoen Wildtuin
Südwestafrika (Namibia)

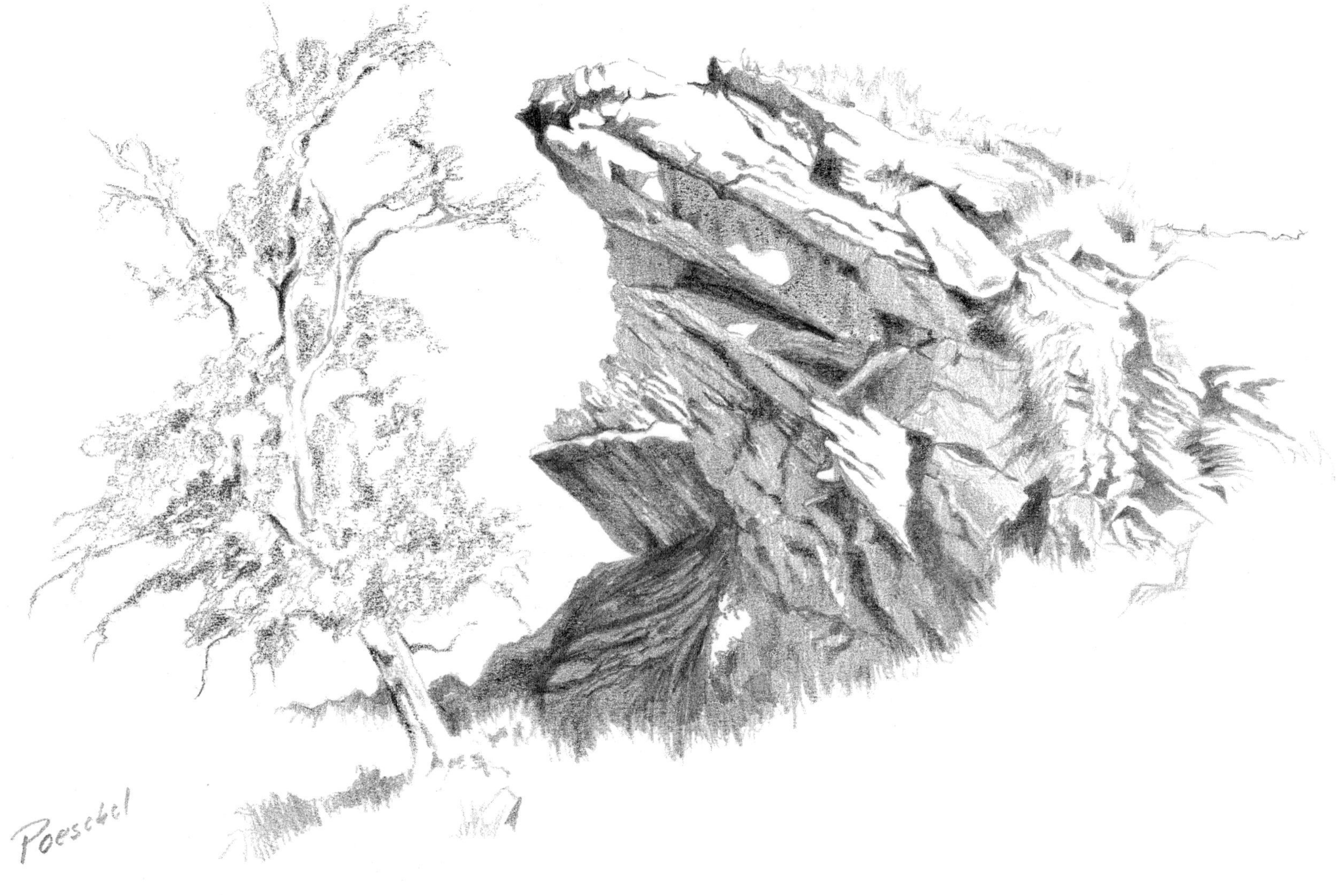
Poeschel

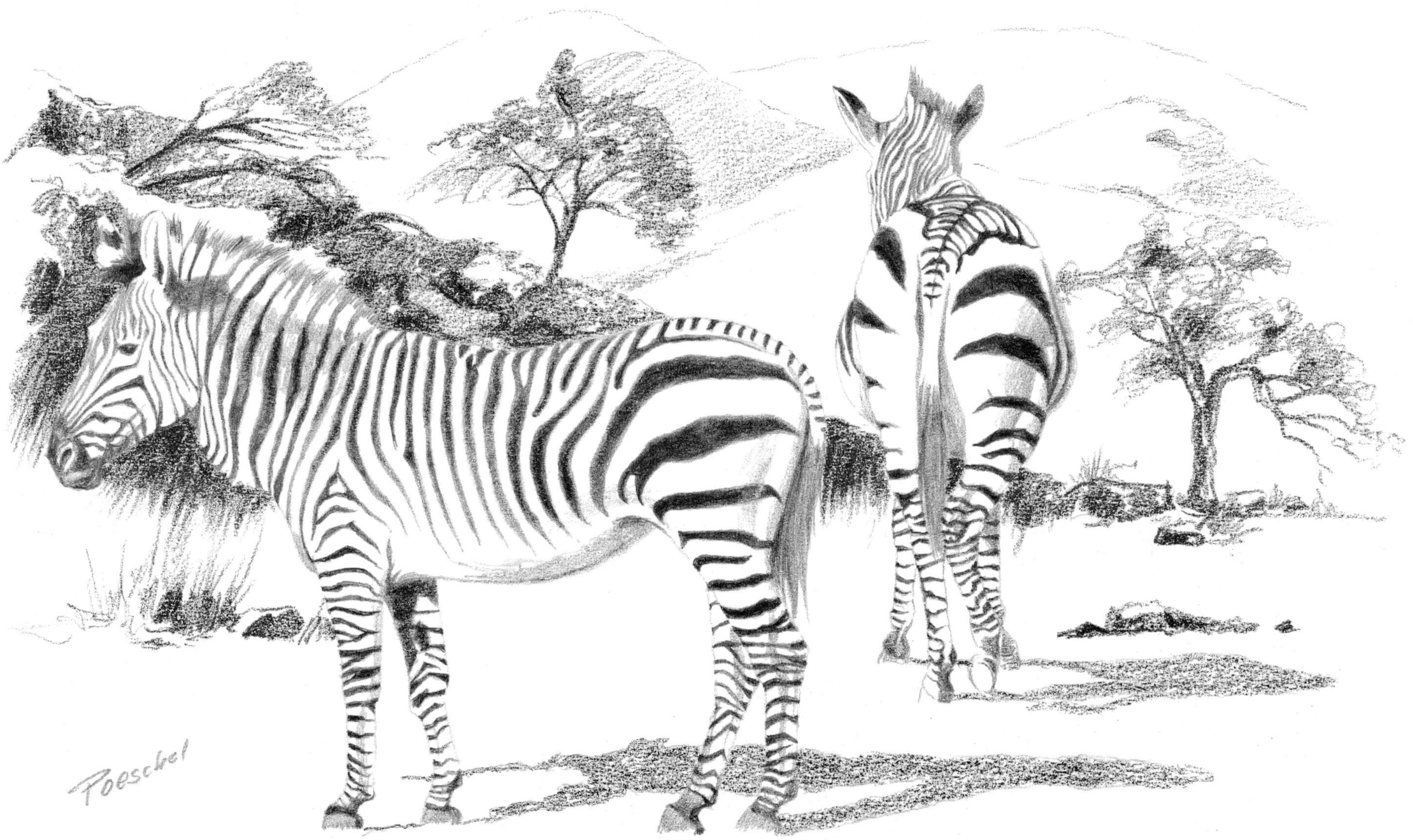
Poeschel

♂

Poeschel

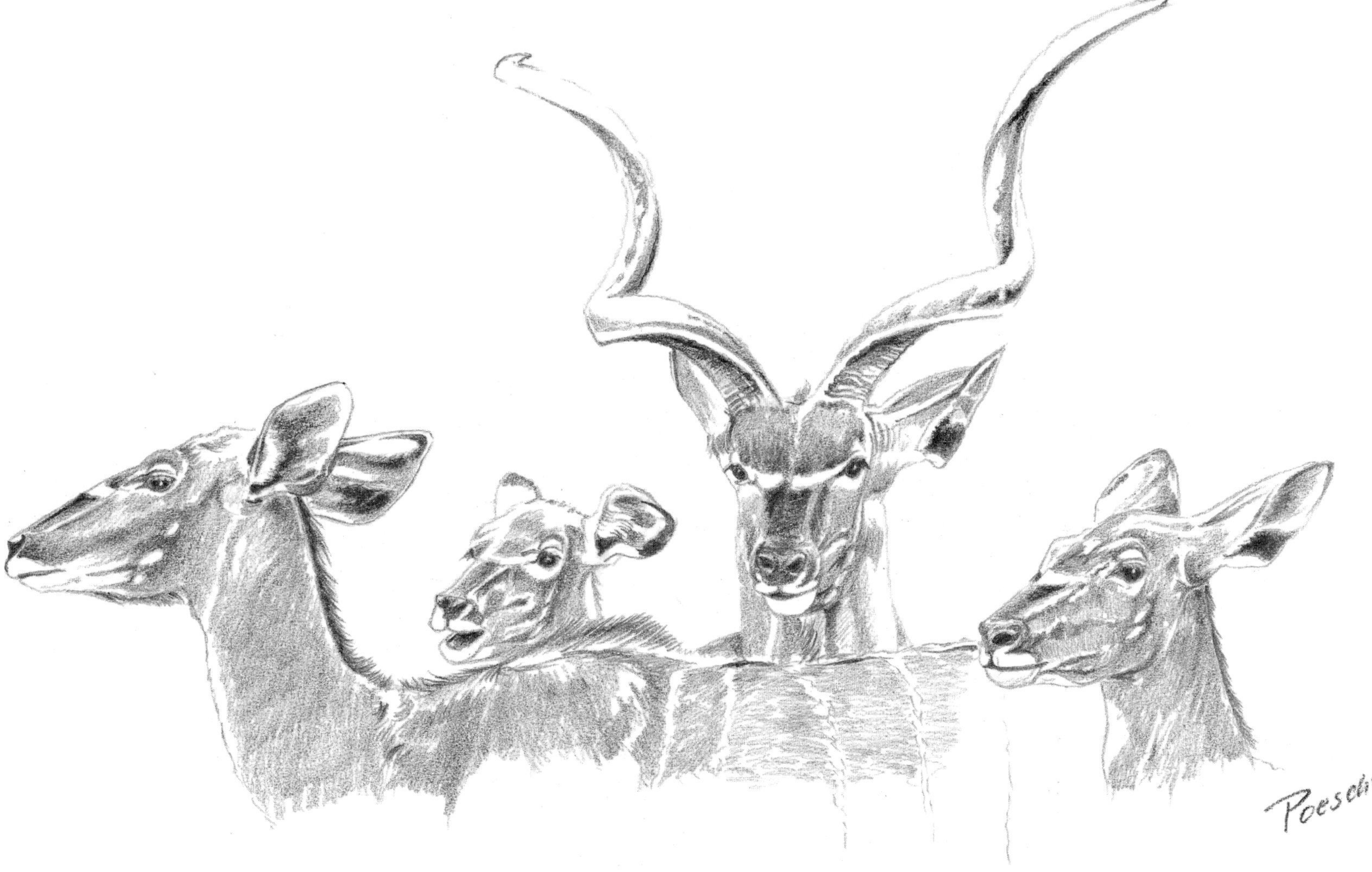
Poeschel

Poeschel

146

Poeschel

Poeschel

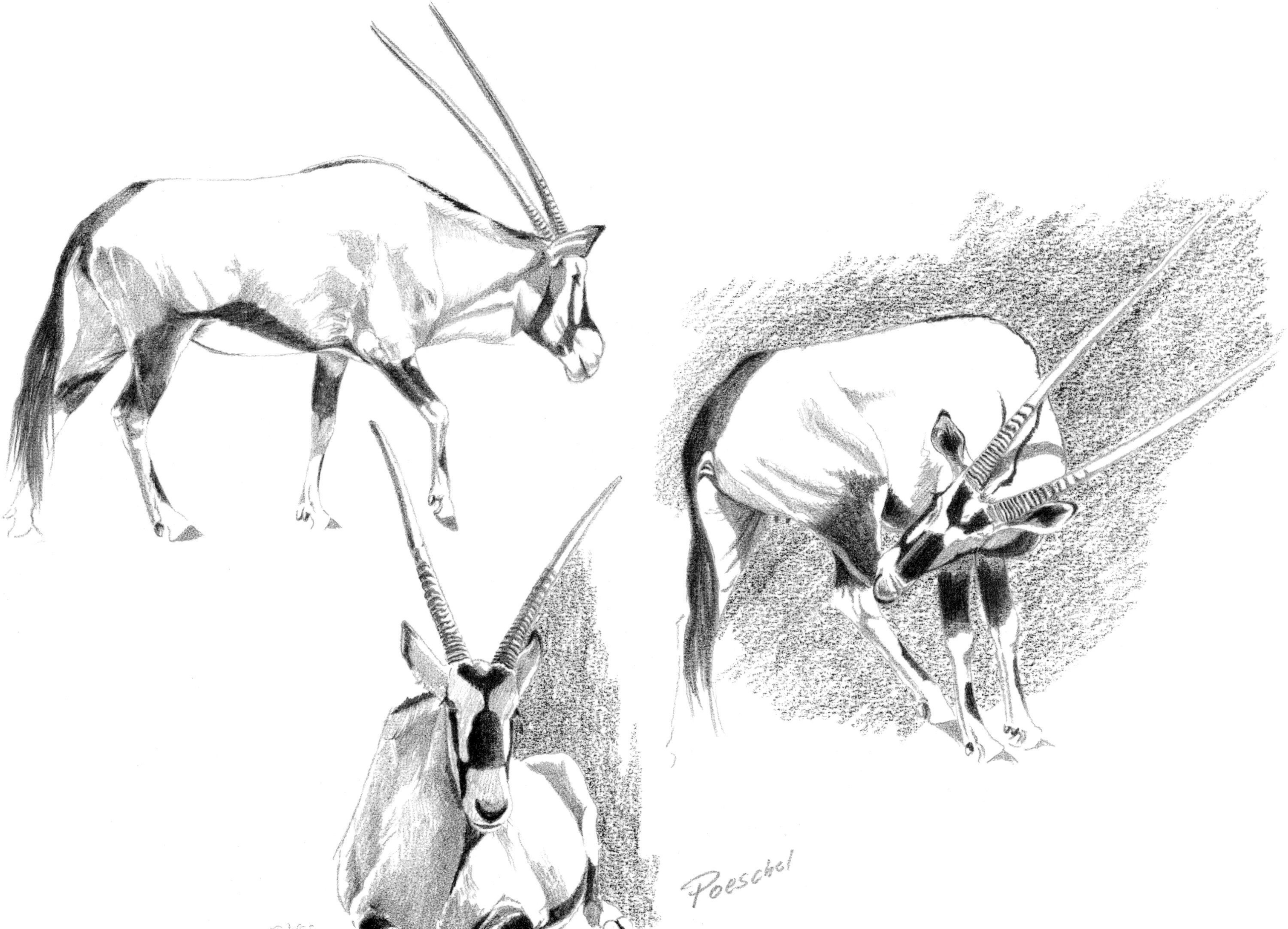
Poeschel

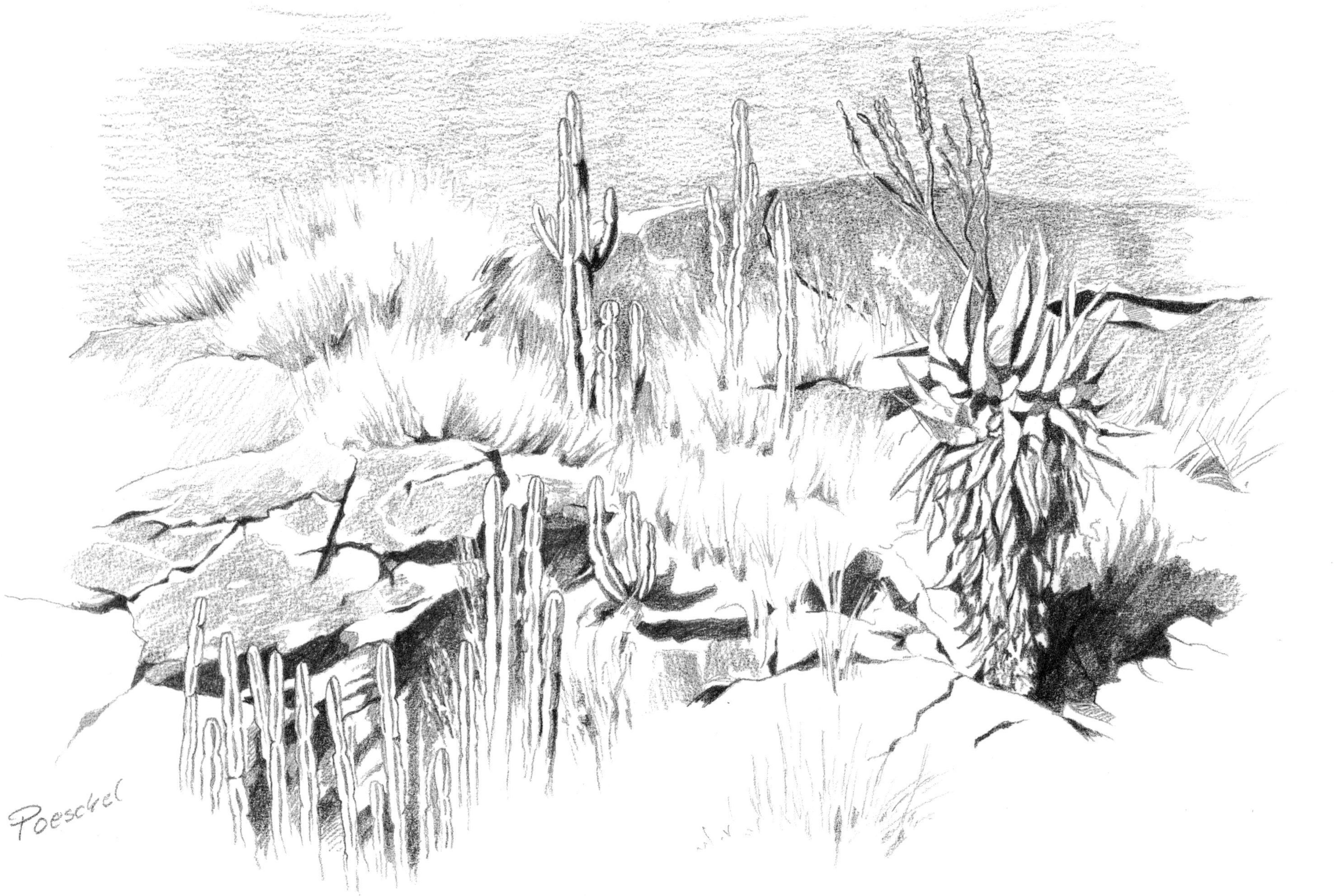
Poeschel

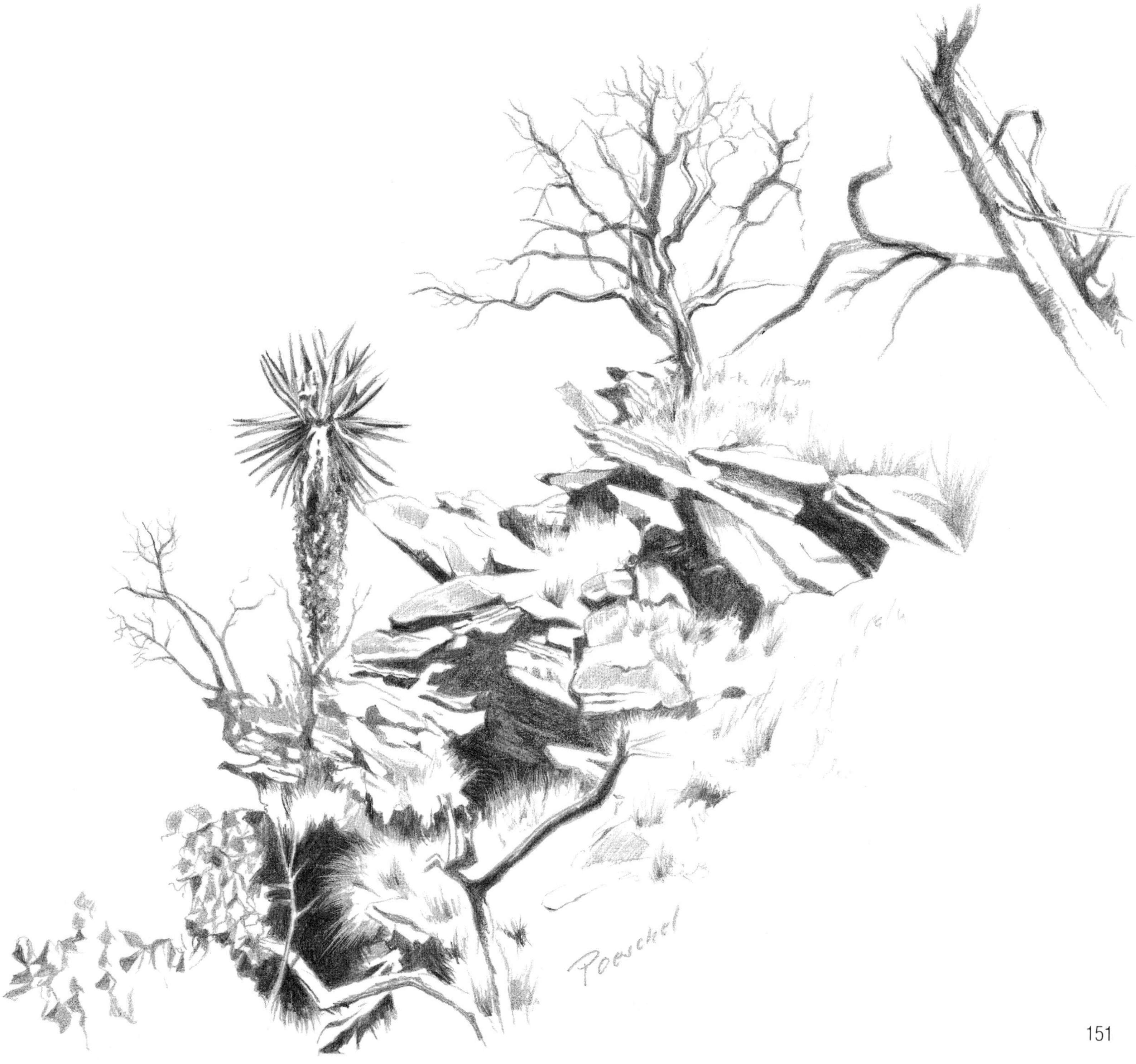

Literatur/literature

Bannister, Anthony and Johnson, Peter
South Africas Wildlife Heritage, Central News Agenca (PTY) LTD

Bechtel, Helmut
Ostafrika in Farbe - Ein Reiseführer für Naturfreunde. - Kosmos, Franck'sche Verlagshandlung 1979, Stuttgart

Davidson, Lynette and Jeppe, Barbara
Acacias. - Centaur Publishers 1981, Johannesburg

Dorst, Jean und Dandelot, Pierre
Säugetiere Afrikas. - Paul Parey Verlag 1970, Hamburg und Berlin

Douglas-Hamilton, Iain und Oria
Unter Elefanten. - R. Pieper & Co. Verlag 1977, München, Zürich

FitzSimons, V.F.M.
Snakes of Southern Africa. - Collins, Reprinted 1978, London and Glasgow

Grandjot, Werner
Reiseführer durch das Pflanzenreich der Tropen. - Kurt Schroeder Verlag 1976, Leichlingen bei Köln

Grzimek, Bernhard und Grzimek, Michael
Serengeti darf nicht sterben. - Ullstein Verlag AG 1959, Darmstadt, Berlin

Grzimek, Bernhard
Grzimeks Tierleben Band 1 bis 13. - Kindler Verlag AG 1971, Zürich

Hagen, Horst
Nationalpark Amboseli. - Kilda-Verlag 1977, Greven

Hagen, Horst
Nationalpark Lake Manyara. - Kilda-Verlag 1978, Greven

Hagen Horst
Nationalpark Serengeti. - Kilda-Verlag 1977, Greven

Hagen Horst
Karibuni-Afrika. - Landbuch-Verlag GMBH 1976, Hannover

Haltenorth, Theodor und Diller, Helmut
Säugetiere Afrikas und Madagaskars. - BLV Verlagsgesellschaft 1977, München, Bern, Wien

Haltenorth Theodor
Klassifikation der Säugetiere: Artiodactyla. - (Handbuch der Zoologie), Walter De Gruiter und Co. 1963, Berlin

Haltenorth, Theodor
Das Großwild der Erde und seine Trophäen. - Bayerischer Landwirtschaftsverlag 1956, Bonn, München, Wien

Hunter, Cynthia
Tsavo Nationalpark. - Ausgabe für deutsche Touristen, Ines May-Publicity, 2. Auflage 1975, Nairobi

Johnson, Peter and Bannister, Anthony
Okawango - Meer im Land, Land im Wasser. - Landbuch Verlag GMBH 1978, Hannover

König, Claus und Ertel, Rainer
Vögel Afrikas Band 1 und 2. - Belser Verlag 1979, Stuttgart und Zürich

Kuhnert, Wilhelm
Im Lande meiner Modelle. - Verlag Klinkhardt und Biermann 1918, Leipzig

Martin, Esmond Bradley
Armes Nashorn: Alle wollen sein Horn. - Das Tier, Zeitschrift vereinigt von Sielmanns Tierwelt, Herausgegeben von Prof. Dr. Bernhard Grzimek und Heinz Sielmann, Heft Nr. 1, Januar 1985

Palmer, Eve
Trees of Southern Africa. - Collins 1977, London and Johannesburg

Petzsch, Hans
Die Katzen. - J. Neumann-Neudamm Verlag 1969, Melsungen, Basel, Wien

Roberts
Birds of South Africa. - The Trustees of the John Voelker Bird Book Fund, C. Struik (PTY) LTD, Fourth Edition 1978, Cape Town

Thomas, Oldfield, Sclater P.L.
The Book of Antelopes Vol. III. - R.H. Porter 1897, London

Tomkinson, Michael
Kenia – Ein Fremdenführer. - Ernest Benn LTD 1978, London

Williams, John G.
Säugetiere und seltene Vögel in den Nationalparks Ostafrikas. - Paul Parey Verlag 1967, Hamburg und Berlin

Williams, John G.
Die Vögel Ost- und Zentralafrikas. - Paul Parey Verlag 1973, Hamburg und Berlin

Zaloumis, E.A. and Cross, Robert
Antelopes of Southern Africa. - Wildlife Society of Southern Africa, Reprinted 1982

Daan-Viljoen-Wildpark. - Herausgeber: Abt. Naturschutz und Fremdenverkehr, S.W.A. Administration 1976

Die Küstenrouten von Natal. - Routenführer für Touristen Nr. 6, Tourist Bureaux, Johannesburg

Lake Nakuru National Park. - Published by the Board of Trustees National Parks of Kenya-an official guide

SAA-tours, South African Airways - Handbuch für Südafrika-Reisen 1980

Süd-West-Afrika. - Informationen für den Besucher, Direktion für Naturschutz und Tourismus, Windhoek

Tanzania National Parks, Lake Manyara National Park, - Guide. - Tanzania Litho LTD, Reprint 1970, Arusha

Touropa Urlaubsberater - Kenia mit Tansania. - Robert Pfützner GmbH 1974, München

Unsere freie Tierwelt - Auf Safari in Südafrika. - satour, South African Tourism Board

AMBOSELI NATIONAL PARK

10 Amboseli-Bulle

13 Der Kilimanjaro (5895 m) bildet den imposanten Hintergrund des Amboseli-Nationalparks

14 Charakteristisch für alle Savannenlandschaften Afrikas sind die Termitenhügel

15 Den überwiegenden Teil des Jahres ist die große Sodafläche des Amboseli-Sees ausgetrocknet

16 Kapitale Kaffernbüffel *(Syncerus caffer caffer)*

17 oben:
Weißrückengeier *(Gyps africanus)*, Sperbergeier *(Gyps rueppeli)* und Marabu *(Leptoptilos crumeniferus)* an einem Flußpferd-Kadaver
unten:
Weißrückengeier *(Gyps africanus)* an den Resten eines Kaffernbüffels
rechts:
Mit Imponiergehabe nähert sich ein Ohrengeier *(Aegypius tracheliotus)*

18 links:
Hammerkopf *(Scopus umbretta)*
rechts:
In der Nähe von Großwild sieht man häufig den Kronenkiebitz *(Vanellus coronatus)*

19 links:
Schwarzhalsreiher *(Ardea melanocephala)*
mitte:
Waffenkiebitz *(Vanellus armatus)*
rechts:
Goliathreiher *(Ardea goliath)* bei den Loginya-Seen

Amboseli-Bull

The Kilimanjaro (5,895 m) forms the impressive background of the Amboseli National Park

All savannah landscapes of Africa are characterized by termite mounds

The vast soda flat of Lake Amboseli is dry during most of the year

African Buffalo

above:
White-backed Vulture, Rueppel's Griffon Vulture and Marabou Stork at the carcass of a hippopotamus
below:
White-backed Vulture at the remains of a buffalo
right:
Lapped-faced Vulture approaches with imposing behaviour

left:
Hammerkop
right:
The Crowned Plover can often be seen around big game

left:
Black-headed Heron
centre:
Blacksmith Plover
right:
Goliath Heron at the Loginya Lakes

20 Im Amboseli-Nationalpark kann man mit Sicherheit den seltenen Geparden *(Acinonyx jubatus)* beobachten

21 Junge Geparden *(Acinonyx jubatus)*

22 An dem Camp bei Ol Tukai stehen große Bestände von Raphia-Palmen *(Raphia australis)*

23 Östliches Weißbartgnu *(Connochaetes taurinus albojubatus)*

24 Der Charakterbaum der ostafrikanischen Savanne ist die Schirmakazie *(Acacia tortilis)*

25 Das Nomadenvolk der Massai ist noch immer sehr traditionsverbunden

TSAVO NATIONAL PARK

27 Elefanten *(Loxodonta africana)* bei den Mzima-Springs

28 Beeindruckende Felsformation in der Nähe der Mzima-Springs

29 Im Vulkangebiet des Tsavo West kann man häufig die zierlichen Klippspringer *(Oreotragus oreotragus aureus)* beobachten

30 Die Impala-Lilie *(Adenium obesum)*, ein stammsukkulentes Hundsgiftgewächs, blüht in der Trockenzeit

31 Die Siedleragame *(Agama agama lionotus)*, scheut nicht die Nähe menschlicher Siedlungen

32 Überall im Gebiet des Tsavo-Nationalparks trifft man auf Elefanten *(Loxodonta africana)*

In the Amboseli National Park the rare Cheetah can certainly be observed

Young Cheetahs

The Campsite of Ol Tukai is surrounded by extensive groves of Raphia palm trees

White-bearded Gnu

The characteristic tree of the East African savannah is the acacia

The Massai are a nomad people whose life is still ruled by strong traditions

Elephants at the Mzima Springs

Impressive rock formations near Mzima Springs

The Klippspringer can often be observed in the volcanic areas of Tsavo West

The Impala-Lily flowers in the dry season

The agame can often be seen near human settlements

Elephants are encountered everywhere in the Tsavo Park

33 12 Uhr Mittags

34 oben:
    Zu den großen Höhepunkten einer Safari in Kenya und Nord-Tanzania gehört der Anblick eines Kleinen Kudu *(Tragelaphus imberbis australis)*
    unten:
    Rotschnabeltoko *(Tockus erythrorhynchus)*
    rechts:
    Gelbkehlfrankolin *(Francolinus leucoscepus)*

35 Brütende Riesentrappe *(Ardeotis kori)*

36 Von besonderer Schönheit und Ästhetik ist die Giraffengazelle *(Litocranius walleri walleri)*

37 Bei Voi im Tsavo-Ost trifft man immer wieder auf starke Kaffernbüffel *(Syncerus caffer caffer)*

38 Der Galana-Fluß kommt durch den Zusammenfluß von Athi und Tsavo zustande. Bei den Lugard-Fällen sind die ausgespülten Felsformationen sehenswert. Im Hintergrund das Yatta-Escarpment

39 Kilimanjaro (5895 m) und Mawenzi (5150 m) vom Mbuyuni-Gate aus gesehen

LAKE NAKURU NATIONAL PARK

41 Der Nakurusee besitzt eine große Kolonie von Weißbrust-Kormoranen *(Phalacrocorax lucidus)*

42 Eine kleine Flußpferdherde *(Hippopotamus amphibius)* lebt am Nordteil des Nakurusees

High noon

above:
The sight of a Lesser Kudu is one of the highlights of a safari in Kenya and North Tanzania
below:
Red-billed Hornbill
right:
Yellow-necked Spurfowl

Kori Bustard breeding

An exceptionally beautiful animal is the Gerenuk

Near Voi in Tsavo East powerful Buffalos are encountered regularly

The joined Athi and Tsavo rivers form the Galana river. Near the Lugard Falls picturesque rock formations carved by water can be seen. In the background is the Yatta escarpment

Kilimanjaro (5,895 m) and Mawenzi (5,150 m), seen from the Mbuyuni Gate

A large colony of White-necked Cormorants exists at Lake Nakuru

A small school of Hippos lives in the northern part of Lake Nakuru

43 links und Mitte oben:
Heiliger Ibis *(Threskiornis aethiopicus)*
Mitte unten und rechts:
Afrikanischer Nimmersatt *(Ibis ibis)*

44 Rosa Pelikane *(Pelecanus onocrotalus)*

45 Siesta bei den Zwerg-Flamingos *(Phoenicopterus minor)*

46 Beim Wasserbock *(Kobus ellipsiprymnus adolfi-friderici)* trägt nur
47 das Männchen ein Gehörn

48 Glanzgans-Erpel *(Sarkidiornis melanotos)*

49 Der Spornkiebitz *(Hoplopterus spinosus)* ist ein Bodenbrüter und versucht
Störenfriede von seinem Gelege durch auffällige Flugmanöver abzulenken

50 Der Riedbock *(Redunca redunca wardi)* ist am Nakurusee immer wieder anzutreffen

51 Im Lake-Nakuru-Nationalpark hatte ich zum ersten Mal auf meinen Reisen
Gelegenheit, den nicht unbedingt seltenen, aber doch schwierig auffindbaren Leoparden zu sehen *(Panthera pardus)*

52 Im lichten Akazienwald beobachtet: Der Perlkauz *(Glaucidium perlatum)*

53 Marabu *(Leptoptilos crumeniferus)*

left and center above:
Sacred Ibis
center below:
Yellow-billed Stork

White Pelican

Lesser Flamingos at siesta

Only the male of the Defassa Waterbuck has horns

Knob-billed Goose

The Spurwing Plover breeds on the ground and tries to avert enemies by conspicuous flight manoeuvres

In the Nakuru Park I saw my first Leopard

The Reedbuck is regularly seen around Lake Nakuru

Detected in acacia trees: the Pearl-spottet Owlet

Marabou

LAKE MANYARA NATIONAL PARK

54 Kaffernbüffel-Schädel

57 Elefantenstudie

58 Affenbrotbaum oder Baobab *(Adansonia digitata)* in der Massaisteppe

59 Am Manyarasee

60 Massaigiraffe *(Giraffa camelopardalis tippelskirchi)* am Seeufer

61 Nur im Manyara- und Ruwenzori-Nationalpark in Uganda klettern Löwen auf Bäume

62 Viele der schönen Schirmakazien *(Acacia tortilis)* sind von Elefanten »skelettiert«

63 Besonders im südlichen Teil des Nationalparks hat man in dem unübersichtlichen Buschgelände überraschende Begegnungen mit Elefanten

64 Die Schwarzfersenantilope oder Impala bevorzugt mit Buschwerk bestandenes Gelände *(Aepyceros melampus rendilis)*

65 Das Spitzlippen-Nashorn ist besser als sein Ruf *(Diceros bicornis bicornis)*

66 Der Anubis-Pavian *(Papio anubis)* bevorzugt erhöhte Aussichtspunkte

67 Nicht häufig ist in Ostafrika der Schmutzgeier *(Neophron percnopterus)*

Skull of Buffalo

Elephant

Baobab in the Massai steppe

At Lake Manyara

Massai Giraffe at the shores of the lake

Lions climb trees only in the Manyara Park and in the Ruwenzori Park in Uganda

Many beautiful acacia trees are stripped by elephants

Especially in the southern part with dense bush, sudden unexpected encounters with elephants can occur

The Impala prefers bushy habitats

Black Rhinoceros

The Anubis Baboon prefers elevated vantage points

Egyptian Vulture, a rare sight in eastern Africa

SERENGETI NATIONAL PARK

69 1,5 Mill. des westlichen Weißbartgnus *(Connochaetes taurinus mearnsi)* leben heute in der Serengeti

70 Charakteristisch für die Serengeti sind die Granit-Kopjes

71 Je nach Jahreszeit sind große Herden des Steppenzebras *(Hippotigris quagga böhmi)* mit anderen Huftieren anzutreffen

72 Von Löwen gerissenes Zebra

73 Kappengeier *(Necrosyrtes monachus)*

74 Zu den Leierantilopen gehört das Topi *(Damaliscus lunatus topi)*

75 Kongonis *(Alcelaphus buselaphus cokii)* sieht man meist in kleinen Trupps. Die lange Schädelform ist besonders auffallend

76 Der Grumeti in der Nordserengeti ist ein ganzjährig Wasser führender Fluß

77 Heute nur noch selten zu sehen ist das Spitzlippen-Nashorn *(Diceros bicornis bicornis)*

78 Immer neugierige Massaigiraffen
79 *(Giraffa camelopardalis tippelskirchi)*

80 Grauhals-Kronenkraniche *(Balearica regulorum)* leben meist paarweise

81 Bei den Thomson-Gazellen *(Gazella thomsoni biedermanni)* hat der Bock ein eigenes Territorium, welches er erbittert gegen Rivalen verteidigt

One and a half million Wildebeest live in the Serengeti

Characteristic features of the Serengeti are the granite Kopjes

Grant's Zebra and other ungulates can be seen in large herds in the Serengeti

Zebra killed by lions

Hooded Vulture

Topi

The Kongoni is normally seen in small parties. Its long skull is a conspicuous character

The Grumeti in the northern Serengeti has water all year

Black Rhino has become rare

Massai-Giraffe

Crowned Cranes live in pairs

The male Thomson's Gazelle has a separate territory which he fiercely defends against rivals

CHOBE NATIONAL PARK

82 Gabelracke *(Coracias caudata)* — Lilac-breasted Roller

85 Fleckenhyänen *(Crocuta crocuta)* — Spotted Hyaena

86 Die Giraffen-Unterart im Gebiet des Chobe ist die Angola-Giraffe *(Giraffa camelopardalis angolensis)* — The Angola Giraffe is the subspecies living in the area of the Chobe

87 Rappenantilopen *(Hippotragus niger niger)* sind sehr wehrhafte Tiere und stellen sich auch Löwen — Sable Antelopes are strong and courageous animals which even fight the lion

88 Das Graufußhörnchen *(Heliosciurus gambianus)* ist ein Baumbewohner — Sun Squirrels live on trees

89 Am alten Ngoma-Gate steht ein riesiger Baobab *(Adansonia digitata)* — A huge baobab near the old Ngoma Gate

90 Über einen langen Zeitraum war der Savuti ohne Wasser und Bäume wuchsen in seinem Flußbett. Doch im Jahre 1957 begann der Fluß wieder Wasser zu führen und 1981, als ich ihn das erste Mal sah, war er erfüllt mit reichem Tierleben — For a long time the Savuti was dry and trees grew in its bed. But in 1957 the river began again to have water, and in 1981, when I saw it for the first time, it was teeming with a rich animal life

91 1983. Verzweifelt gruben Elefantenbullen im ausgetrockneten Flußbett nach Wasser. Auch Südafrika wird seit Jahren von einer großen Trockenheit heimgesucht — 1983. Elephant bulls dug desperately for water in the river bed. South Africa, too, has been haunted for years by terrible droughts

92 Das Camp am Savuti wird regelmäßig von einem großen Elefantenbullen besucht — The Savuti Camp is regularly visited by a huge elephant bull

93 Während unseres Aufenthaltes am Savuti waren ein Pärchen Gelbschnabeltokos *(Tockus flavirostris leucomelas)* unsere Campgenossen — A pair of Yellow-billed Hornbills were our companions during our stay at the Savuti

94 Ein besonderes Erlebnis: Von Löwen gerissener Kaffernbüffel mit einem
95 Rudel aggressiver Hyänen — Buffalo killed by lions, with a gang of aggressive Spotted Hyaenas

96 links:
Der Gaukler *(Terathopius ecaudatus)* ist durch seinen kurzen Stoß gut von anderen Greifen zu unterscheiden
rechts:
Den Sekretär *(Sagittarius serpentarius)* findet man in allen offenen Landschaften und Savannen Afrikas. Er ist an dem langen Schwanz und dem hängenden Federschopf am Hinterkopf leicht zu erkennen

97 In der Weite der Savuti-Marsh

MOREMI WILDLIFE RESERVE

99 Seit dem Jahre 1975 ist die Jagd auf Krokodile *(Crocodylus niloticus)* wegen Unergiebigkeit durch die Regierung untersagt. Die Bestände erholen sich wieder langsam

100 Das aus einem Stamm gehauene »Mokoro« ist das Verkehrsmittel der Eingeborenen im Okawango-Delta

101 Die Erhabenheit und Stille der Natur ist ein ungeheurer Kontrast zu der Hektik und dem Lärm in Europa

102 Ein territorialer Flußpferdbulle *(Hippopotamus amphibius)* gibt uns deutlich zu verstehen, daß ihm an unserem Besuch nicht gelegen ist

103 Die Brücken im Moremi Wildlife Reserve sind aus dem eisenharten Mopaneholz

104 Zum ersten Mal hatte ich hier die Gelegenheit, ein Rudel Wildhunde *(Lycaon pictus pictus)* aus unmittelbarer Nähe zu beobachten

left:
Bateleur, characterized by a very short tail

right:
The Secretary Bird is found in all open landscapes of Africa. It can be identified by its long tail and the long feathers hanging from the back of its head

In the expanse of the Savuti marshes

Since 1975 the hunting of crocodiles has been prohibited by the government. The populations are regenerating slowly

The mokoro, a dugout canoe, is the transport vehicle of the natives in the Okawango delta

Majesty and silence of nature are a tremendous contrast to the hectic activity and noise in Europe

A territorial Hippo bull lets us know clearly that he doesn't care for our visit

The bridges in the Moremi Wildlife Reserve are made of mopane-wood, which is hard as iron

For the first time here I could watch a pack of Wild Dogs at close quarters

105 Die für die seichten Überschwemmungsgebiete des Okawango-Delta typische Antilope ist die Litschi-Moorantilope *(Kobus leche leche)*

106 Sporngänse *(Plectropterus gambensis)* trifft man paarweise oder auch in größeren Ansammlungen

107 links:
Zur Familie der Bienenfresser gehört der Zwergspint *(Merops pusillus)*
rechts:
Klunkerkranich *(Grus carunculatus)*

108 Mopane-Baum mit Webervogelnestern

109 Sonnenuntergang am Xuroro-Camp

KRUGER NATIONAL PARK

110 links oben:
Vor dem Eingang zum Paul Kruger Gate erinnert eine Büste an den Burenpräsidenten und Gründer des nach ihm benannten Nationalparks
unten und rechts:
Überall im Nationalpark trifft man auf Impalas *(Aepyceros melampus melampus)*, welche mit einer Kopfzahl von 153 000 die größte Tiergruppe darstellen

113 Nirgendwo sonst konnte ich den Kaffernhornraben *(Bucorvus leadbeateri)* so eingehend beobachten, wie in der Nähe vom Skukuza-Camp

114 Granit-Inselberg im südlichen Teil

The Lechwe Antelope is typical of the shallow flooded parts of the Okawango delta

The Spur-winged Goose can be seen in pairs or in bigger flocks

left:
Little Bee-eater
right:
Wattled Crane

Mopane tree with weaverbird nests

Sunset at the Xuroro Campsite

left above:
In front of the entrance to the Kruger Gate a bust memorializes the Boer president and founder of this national park
below and right:
Everywhere in the park the Impala, the most numerous species (153,000 animals) can be seen

Nowhere else could I watch the Ground Hornbill so intensively as in the vicinity of the Skukuza Camp

Granite mountain in the southern part

115 Der Pavian Südafrikas ist der Tschakma- oder Bärenpavian *(Papio ursinus)*

116 Weiblicher Großer Kudu *(Tragelaphus strepsiceros strepsiceros)* mit Rotschnabel-Madenhackern *(Buphagus erythrorhynchus)*

117 Beim Großen Kudu *(Tragelaphus strepsiceros strepsiceros)*, trägt nur das Männchen ein spiralig gewundenes Gehörn

118 Der flugunfähige Strauß *(Struthio camelus australis)* ist an das Leben in der Savanne hervorragend angepaßt

119 links:
Der Graulärmvogel *(Corythaixoides concolor)* hat mit seinem »gwäh, gwäh« (= engl. »go-away«) schon manchem Jäger die Jagd verdorben
rechts:
Nacktkehl-Frankolin *(Francolinus afer lehmanni)*

120 Den überwiegenden Teil des Tages verbringen Löwen
121 *(Panthera leo krugeri)* schlafend oder ruhend irgendwo im Schatten

122 Das Gebiet am Sabie-River wurde als Sabie Game Reserve 1898 von Präsident Paul Kruger, als erster Abschnitt des heutigen Kruger-Nationalparks unter Schutz gestellt

123 Zwei kuriose Tiergestalten
links:
Elefantenspitzmaus *(Elephantulus rufescens)*
rechts:
Warzenschwein *(Phacochoerus aethiopicus aeliani)*

124 Einer der großen Augenblicke im Kruger-Nationalpark war für mich der Anblick eines Karakals *(Lynx caracal limpopoensis)*, der, da nachtaktiv, bei Tag nur selten zu sehen ist

125 Zierlich und anmutig, das Steinböckchen *(Raphicerus campestris capricornis)*

The Chacma Baboon is the baboon species living in South Africa

Cow of Greater Kudu with Red-billed Oxpeckers

Only the male Greater Kudu has the spirally twisted horns

The flightless Ostrich is well adapted to life in the savannah

left:
The Go-away-Bird is a great nuisance to hunters; it warns the game with noisy calls
right:
Red-necked Francolin

The Lions spend most of the day resting or sleeping somewhere in the shade

The area around the Sabie river, declared the Sabie Game Reserve by president Kruger in 1898, was the first part of the Kruger National Park brougth into protection

Two extraordinary animals
left:
Elephant Shrew
right:
Warthog

One of the highlights of my visit to the Kruger Park was the sight of the Caracal, which is rarely seen at daytime

Delicate and graceful, the Steenbok

HLUHLUWE GAME RESERVE

127 Dem strengen Schutz der südafrikanischen Naturschutzbehörden ist es zu
128 danken, daß das Breitlippen-Nashorn *(Ceratotherium simum simum)* nicht
129 nur vor dem Aussterben bewahrt, sondern wieder in Gebiete eingebürgert werden konnte, in welchen es ausgerottet war.

130 links:
Blüte des Korallenbaumes *(Erythrina indica)*
rechts:
Paradiesvogelblume *(Strelitzia reginae)*

131 Blick vom Restcamp auf die Hügellandschaft Natals

132 Die Nyala-Antilope *(Tragelaphus angasi angasi)* hat ein kleines Verbreitungs-
133 gebiet. In den Wildschutzgebieten Natals kann man sie mit Sicherheit beobachten. Besonders die männlichen Tiere sind sehr auffällig gezeichnet

134 links und Mitte:
Den Kronentoko *(Tockus alboterminiatus)* konnte ich über einen längeren Zeitraum beobachten.
rechts:
Der Raubadler *(Aquila rapax)* ist in Afrika weitverbreitet

135 Überall in nicht zu offenem Gelände sieht man die Grüne Meerkatze *(Cercopithecus aethiops pygerythrus)*

136 Auch das Hluhluwe Game Reserve besitzt eine Population des fast ausgerotteten Burchell-Zebra *(Hippotigris quagga burchelli)*

137 Das Natal-Warzenschwein *(Phacochoerus aethiopicus sundevalli)* ist heute die südlichste Unterart

Due to the strict protection by the South African nature conservation authorities the White Rhino has not only been saved from extincton, but could again be repatriated in areas where it had been extirpated before

left:
Coral Tree
right:
Crane Flower

View from the Rest Camp towards the hilly landscape of Natal

The Nyala Antelope has a restricted distribution. It can be seen for certain in the game reserves of Natal. The males have an especially conspicuous coloration

left and center:
Crowned Hornbill

right:
The Tawny Eagle has a wide distribution all over Africa

Everywhere in not too open terrain the Vervet Monkey can be found

A population of the nearly extinct Burchell's Zebra also survives in the Hluhluwe Game Reserve

Today the southernmost subspecies of the Warthog lives in Natal

DAAN VILJOEN WILDTUIN

138 links:
Eines der Wahrzeichen Windhuks, der Hauptstadt von Südwestafrika (Namibia), ist das Reiterdenkmal, welches zu Ehren der Angehörigen der deutschen Schutztruppe errichtet wurde, die während der Herero- und Namafeldzüge in den Jahren 1904 bis 1908 gefallen sind. Von dem Berliner Bildhauer Adolf Kürle ausgeführt, wurde es am 27. Januar 1912 zu Kaisers Geburtstag enthüllt.
rechts:
Ecke Kaiser- und Lüderitzstraße in Windhuk steht das bekannte Kudu-Denkmal, welches die Richtung nach Norden zum berühmten Etoscha-Nationalpark anzeigt. Es wurde von Prof. Fritz Behn geschaffen.

141 In Südwestafrika, vorwiegend auf Farmgelände, ist der Große Kudu *(Tragelaphus strepsiceros strepsiceros)* noch sehr zahlreich

142 Unerschöpflich für den Künstler sind die Landschaftsmotive im Daan Viljoen Wildtuin

143 Das Hartmann-Bergzebra *(Hippotigris zebra hartmannae)* war schon fast ausgerottet, doch durch strenge Schutzmaßnahmen haben sich die Bestände wieder erholt. Auch im Daan Viljeon Wildtuin lebt eine kleine Population

144 Die Elen-Antilope *(Taurotragus oryx oryx)* ist die schwerste aller Antilopen, sie kann bis 1000 kg schwer werden

145 Den Großen Kudu *(Tragelaphus strepsiceros strepsiceros)* trifft man meist in kleineren Verbänden

146 Der Charakterbaum Südwestafrikas ist der Kameldornbaum *(Acacia giraffae)*
links:
Schoten des Kameldornbaumes

left:
One of the »landmarks« of Windhoek, the capital of Southwest Africa (Namibia), is the equestrian statue, which was set up in honour of those members of the German colonial troops, killed during the Herero and Nama rebellions between 1904 and 1908

right:
At the corner of Kaiser- and Lüderitz-Street there is the well-known Kudu Statue, which shows the way towards the Etosha National Park

In Southwest Africa, especially on farmland, the Greater Kudu is still quite numerous

The landscape motifs of Daan Viljoen Wildtuin are an inexhaustible source for the artist

The Mountain Zebra had been near extinction; through strict protection measures its numbers have increased again. A small population lives in the Daan Viljoen Wildtuin

The Eland is the largest antelope, it can weigh up to 1000 kg

The Greater Kudu is usually encountered in small groups

The Camel thorn is a characteristic tree of Southwest Africa

left:
pods of the Camel thorn

147 Im Daan Viljoen Wildtuin traf ich auf eine große Artenvielfalt in der Vogelwelt
links:
Akaziendrossel *(Turdus litsipsirupa)*
rechts oben:
Maskenbülbül *(Pycnonotus nigricans)*
rechts unten:
Rotschnabelfrankolin *(Francolinus adspersus)*

148 Der südlichste Vertreter der Kuhantilope ist das Kaama *(Alcelaphus buselaphus caama)*

149 An die Trocken- und Wüstengebiete hat sich der Spießbock *(Oryx gazella gazella)* hervorragend angepaßt

150 Zwei charakteristische Landschaftsmotive aus dem über 2000 m hohen
151 Khomas-Hochland

In the Daan Viljoen Wildtuin I observed a great variety of bird species
left:
Ground-scarper Thrush
right above:
Red-eyed Bulbul
right below:
Red-billed Francolin

Red Hartebeest, the southernmost representative of the hartebeest

The Gemsbock is well adapted to dry habitats and desert areas

Two characteristic landscape motifs in the Khomas highlands at an elevation of more than 2000 meters